A CONCISE GUIDE
FOR WRITERS

A CONCISE GUIDE
FOR WRITERS

SECOND EDITION

Louis E. Glorfeld
University of Denver

David A. Lauerman
Canisius College, New York

Norman C. Stageberg
University of Northern Iowa

HOLT, RINEHART AND WINSTON, INC.

New York Chicago San Francisco Atlanta
Dallas Montreal Toronto London Sydney

ACKNOWLEDGMENTS

The authors are grateful to Dr. Andrew MacLeish of the University of Minnesota for his assistance with "Overcoming High-Frequency Faults," and to their many freshman students at the University of Northern Iowa and Northern Illinois University for testing these materials in prepublication form.

Printed in the United States of America

90123 9 9876543

PREFACE

This brief textbook has four purposes: (1) to show college freshmen how to plan and begin a theme; (2) to help them overcome those writing faults that seem to occur most frequently in freshman composition; (3) to teach them the standard methods of developing thought in paragraphs and the devices that will enable them to achieve a smooth, readable style; and (4) to show them how to cope with an essay examination.

The first section offers the student practical hints on getting started and suggests how to draw up a quick scratch outline for short themes.

The second section takes up the more common writing faults one at a time. A brief and simple explanation is given for each fault, with ample illustrations drawn from student writing. Meticulous qualifications and overrefinements are avoided. Following the explanation, two revision exercises are presented. Each consists of sentences taken mostly from themes written by freshmen and embodies the fault with which the explanation is concerned. Thus the student, by grappling with real, not factitious, writing problems, has the sense of performing a task that will bring a welcome reward—success in the mastery of one important step toward sound writing.

The third section deals with two broader problems which plague the beginning writer: thought development and coherence. It also includes a unit on the student's most onerous task —writing an essay examination—and concludes with a treatment of an outline for a long theme or paper.

It should be understood that this guidebook is not a grammar or a complete handbook of composition. Rather, it represents a thoughtful selection of the most apparent problems garnered from the maze of material found in the conventional freshman text. The authors have learned, through using the guide in the classroom, that its economy of presentation is helpful in clarifying difficulties for the student.

The instructor can use the guide in several ways: (1) he may discuss the various writing faults and assign the revision exercises in the order in which they occur, (2) he may assign the pages dealing with those faults with which the class as a whole is having trouble, or (3) he may make remedial assignments for individual students.

The student may refer to the explanations and exercises to help himself in dealing with the specific faults which recur in his writing.

L. E. G.
D. A. L.
N. C. S.

Denver, Colorado
Buffalo, New York
Cedar Falls, Iowa
November 1968

CONTENTS

PREFACE   v

1   Beginning to Write   1

Getting Under Way   1

2   Overcoming High-Frequency Faults   9

1.	Agr SV Agreement of Subject and Verb	9
2.	Amb Ambiguity	12
3.	AP Use of Active Instead of Passive Recommended	15
4.	Cap and LC Capitalization and Lower Case	18
5.	CC Comma or Commas for Clearness	21
6.	CF Comma Fault or Comma Splice	23
7.	C Uses of the Comma	26
8.	Cl Cliché	30
9.	Col Use of Colon Recommended	32
10.	Colloq Colloquial	34
11.	Conf Confused Sentence	35
12.	Dash Use of Dash Recommended	38
13.	Frag Fragment	40
14.	Imp Imprecision	42
15.	M Meaning Is Not Clear	44
16.	Mis Misplaced Modifier	46

17. Mis OM   Misrelated Opening Modifier   48
18. MM   Mixed Metaphor   50
19. NS   Nonstandard   52
20. OBS   Overburdened Sentence   53
21. ¶/no ¶   Paragraph Divisions   56
22. PAC   Punctuation of Adjective Clauses   60
23. Paral   Parallel Structure   63
24. Per   Periodic Structure   65
25. Poss   Possessive   68
26. Quot   Mechanics of Quotation   71
27. Red   Redundancy   75
28. Ref   Reference of Pronouns   76
29. Rep   Repetition   79
30. RO   Run-on (Fused) Sentence   81
31. SB   Sentence Beginning   83
32. S Col   Use of Semicolon Recommended   86
33. Sl   Slang   89
34. SP   Shift of Person   90
35. Sp   Spelling   92
36. Sub   Subordination   95
37. SV   Sentence Variety   97
38. T   Transition   99
39. Te S   Tense Shift   101
40. W   Wordiness   104
41. WW   Wrong Word   106

3 Mastering Larger Writing Problems

3 Mastering Larger Writing Problems   112

1. Achieving Sentence Flow   112
2. Developing Thought in Paragraphs   126
3. Writing Essay Examinations   144
4. Organizing by Outline   165

INDEX   179
CORRECTION CHART   184

1

BEGINNING
TO WRITE

Getting Under Way

Writing clean and simple prose is difficult, and many writers
with years of experience still find it a slow and painful process.
So if you find the going heavy and laborious, if you fumble for
words and get lost in labyrinthine sentences, and if you cannot
work your way out—do not despair. Everyone who writes has
this trouble.

The hardest part of writing is the beginning. You may find
yourself just sitting there and staring into space, unable to
get under way. This is what is known as "white-paper paral-
ysis." If this happens to you, here is a procedure that some
writers find helpful: (1) Jot down quickly whatever comes to
mind as you consider the topic you have been assigned. Even
if your first notations seem trivial and uninspired, they will
undoubtedly awaken further words and ideas. (2) Organize
your jottings into groups that would seem to go together. Re-

1

ject those which do not fit anywhere or which seem unimportant to your purpose. Each group can become a division of your paper. If you have too many groups, however, the result would be a number of very short paragraphs, so try regrouping into larger sections. (3) Arrange the groups in some sensible order. Now you have what is called a scratch outline, and you are warmed up and ready to begin. (4) Write the first draft rapidly, without bothering about details of spelling and punctuation, to get your thoughts into some sort of form. When you reach this stage, you are out of the woods.

Let us go through this process again in more detail, using examples. Some people call this first step "brainstorming." Remember that at this stage you do not know exactly what you are going to write. Relax—anything goes. Suppose you have been asked to write a short theme on "My Summer Job." Do not ask, "What will I say about it?" Instead, start jotting down a list of words or sentences—your scrap pile of ideas, anything you *might* want to use. A very short list could look like this:

City Park Department
cutting grass
lawnmowers (and sickles)
the crew: Ed, George, and Me
ball games
swimming pool
the boss, Red
Old John, the caretaker
grooming the baseball diamond
girl-watching

You have ten points listed; if three of them prove to be major points from which you can build individual paragraphs, and if three of them can be used to develop each of those paragraphs, you are ready to start writing.

Sorting through your list, you note that any one of the items could serve as the topic for a short paragraph. But that is not what you are looking for; you want an orderly two- or three-paragraph theme. So make your choices: which of these notes or groupings of notes are inclusive enough to serve as the topics of your two or three paragraphs? "The crew" is an obvious topic. The parallel *-ing* verb forms (*cutting, grooming*) indicate a second topic: "the kinds of work we did." The remaining items seem hard to manage ("ball games," "girl-watching"), since they were only diversions or distractions from the job. Well, there we have the third topic: "Diversions"!

Here is the result of your first sorting:

THE CREW:	Ed, George, and Me
	Red, the boss
	Old John, the caretaker
THE KINDS OF WORK:	cutting grass
	grooming the baseball diamond
DIVERSIONS:	games
	the swimming pool
	girl-watching

It becomes obvious as you consider these groupings that they need rearranging, rephrasing, and filling-in, but note that already two important things have been done. First, you have decided what to talk about by grouping your material around three major topics—the crew, the kinds of work, and the diversions. Second, you have left "City Park Department" and "lawnmowers and sickles" unused, thereby setting aside material that seems of secondary importance.

It is time to rearrange your line-up and fill in the gaps. Again, you find that you are farther along than you had suspected. Without noticing how or why, you have picked out the simplest and most immediate topic, "the crew," to work with first. Now, if you see that as an easy way into your topic, very likely your reader will too. Similarly, you have left the most engaging topic for the last. This is the position where you make the final impression on the reader. The sober business of the park job itself is left where it belongs, in the first two paragraphs. This seems a sensible approach to the overall arrangement of your theme.

Now for another look at the arrangement of items within these groupings. The first topic seems satisfactory:

```
THE CREW:    Ed. George, and Me
             Red, the boss
             Old John, the caretaker
```

You may want to put "Red" first, though at this point the order does not seem important. The second topic looks thin for the middle paragraph; only two points are listed:

```
THE KINDS OF WORK:    cutting grass
                      grooming the baseball diamond
```

You might add the tools, "lawnmower and sickle," or look for other material, such as "cleaning the swimming pool." Or you might plan to illustrate the main jobs more fully, in this way:

```
THE KINDS OF WORK:    cutting grass—
                          edging behind the tractor mowers
                          sickling the grass along the fences
                          planting new grass in bare spots
                      grooming the baseball diamond—
                          raking and rolling the infield
                          marking the foul lines and setting the
                          bases
                          sweeping the grandstands
```

The third topic also needs attention:

```
DIVERSIONS:    ball games
               the swimming pool
               girl-watching
```

For one thing, you want to specify that *watching* the ball games and *watching the people* at the swimming pool is what you have in mind. They go nicely with "girl-watching," which would make a pleasing conclusion and might even deserve a separate paragraph.

You now have a scratch outline. The process which has taken us several paragraphs to describe is one which you will ordinarily perform in a matter of several minutes of note-making, sorting, and rearranging. If you took the trouble to write it out neatly, it would look like this:

```
THE CREW:    Ed, George, and Me
             Red, the boss
             Old John, the caretaker
```

THE KINDS OF WORK: cutting grass—
                                    edging behind the tractor-mower
                                    sickling tall grass along the fence
                                    planting new grass in bare spots
                                    grooming the ball diamond—
                                    raking and rolling the infield
                                    marking foul lines and setting bases
                                    sweeping the grandstands

DIVERSIONS: watching ball games
                            lounging near the swimming pool and watching the people
                            girl-watching

To stop now for so much as a drink of water might give all this a chance to dribble away. While it is clear in your mind, you must pour out the words of the first draft. So, with the reservation that what you write at this point will be revised, stake out a tentative first sentence, and watch the rest follow:

I spent last summer working with a City Park Department crew. The pay was not much, but neither was the work, and the guys on the crew had a ball together. There were five of us on the crew. Ed, George, and I were the hardcore goof-offs. Ed spent his time watching girls, George talked incessantly about baseball, and I sort of tagged along. None of us worried much about the job except Red, the boss. He worked for the city year-round, so he had to. So did Old John, the caretaker, but for some reason he came and went as he pleased.

Our job was mainly cutting grass and grooming the baseball diamond. Every Wednesday the Toro tractor-mower came around, and we followed behind it with hand-mowers and sickles, trimming and edging. We used the sickles on the tall grass along the fences, which was not only back-breaking but boring, since we usually worked alone at this. In between times we planted new grass or sod in bare spots. Grooming the baseball diamond was more fun, and a lot easier. Before each game we raked and rolled the infield before marking the foul lines and batter's box and setting out the bases. That always gave George a chance to stride to the mound and demonstrate his "rocking motion" and "cautious glance at the runner on first." Ed preferred sweeping the peanuts and candy-wrappers out of the grandstand; that gave him a chance to lean on his broom and tell about the date he had almost had.

By this time, the serious problem of "getting started" is behind you. The last paragraph will come easily, now that you have your material in mind and working for you.

After this first flurry of composition, much polishing still remains to be done. You will have to revise and re-revise, rearrange words and shift groups, erase and add, tidy up your spelling and punctuation, and consult your dictionary. In this process you will perhaps make your language less colloquial, for such terms as "goof-off" and "sort of" hardly seem appropriate in written style. When you have finished the rewriting and the result looks good, set your paper aside to cool. Next day reread it. You may be surprised to see how it has deteriorated, but you will recognize the flaws easily before you copy the final version.

The real trick to getting started, and the one you need to practice, therefore, is this basic process of first "brainstorming," then sorting out, and finally arranging the groups of material on the topic assigned.

A few more examples may be helpful at this point. Here is a set of theme topics of a sort often handed out to freshmen at the beginning of the fall semester:

Freshman Orientation

My First Week on Campus

The Difference between High School and College

Why People Like the Beatles

Why I Came to College

Which one would you choose as the topic for a short theme? For the first two you need to have some definite impressions in mind; if you have them, either would be a good choice. The last three direct you to a specific approach to the topic given; they require you to offer examples or reasons to support what you say. You will choose one of these if some examples or reasons come to mind, and if you feel that you can think of others.

Suppose we brainstorm one of the two descriptive topics— "My First Week on Campus." All you need do here is ask yourself, "What happened?" and begin your list of notes:

finding my room
unpacking
the orientation meetings
mixers
registration for classes
finding my way around campus
getting lost downtown
my roommate
my first class

That was a full week, and you may find it hard to stop making notes. But you already have several key points, and can begin arranging them. One method would be to arrange the points chronologically, beginning with your arrival and ending with your first class. Another would be to list the types of activities: "Orientation," "Getting ready for classes," "Meeting people," "Finding my way around." With this set of headings you would prune away subjects like "unpacking" that are of only personal interest. Now you can go ahead and arrange your scratch outline.

A different approach is needed for a topic such as "Why I Came to College." In this case ask yourself "What were my reasons?"

to continue my education
to prepare myself for a job
because I did not want to be a drop-out
my parents and teachers advised me to
I wanted to get away from home
my friends were all going to State
I am too young to get married
I have not yet decided on a career

You have several points listed here, but as you look them over you may note that two are not consistent. For example, "to prepare myself for a job" and "I have not yet decided on a career" seem contradictory, so one must go. You also must establish your primary reasons for coming to college. That requires a bit of thinking, and only you can identify the reasons behind your choice. But the choice is crucial for the development of your theme; it will shape the paragraphs you write and the conclusions you draw.

Happily, most students face the problem of getting started only in writing their first few themes. As you learn more about organization and the development of ideas (*see* "Developing Thought in Paragraphs"), this procedure will become a matter of habit, and you will no longer need to fear a blank sheet of white paper.

EXERCISES

1. Write the missing final paragraph for "My Summer Job." You will, of course, have to imagine the details. Prepare yourself by rereading the first two paragraphs and the scratch outline. For an opening sentence, you might try "As you see, the job was boring, so we looked for diversions," or "The rest of our time we spent girl-watching."
2. Arrange the notes on "My First Week on Campus" under the four headings suggested (types of activities), and finish the scratch outline. Add details to fill out, if you wish.
3. Sort out the notes on "Why I Came to College" and arrange them into a scratch outline. Add any details that would be helpful.
4. Produce a set of notes and a scratch outline for one of the other three topics in the list on page 6.

2

OVERCOMING HIGH-FREQUENCY FAULTS

1 Agr SV

Agreement of Subject and Verb

Most nouns and pronouns in English are either singular or plural in form. For example, these are singular: *duty, thing, each, it, neither;* whereas these are plural: *duties, things, both, they, many.* Similarly the verb (in the present tense, third person) has an *-s* form to denote the singular. Here are some singular forms of the verb: *is, was, has, does, depends, chooses, varies;* and the parallel forms without the *-s* are plural: *are, were, have, do, depend, choose, vary.*

What is important here is that the singular form of the noun must be matched with a singular form of the verb, as in these: *neither is, a thing was, the thickness varies, each has.* Likewise a plural noun requires a plural verb, that is, a verb without the *-s*, as in these combinations: *purposes are, things have, duties depend.* All this is elementary and you certainly know enough not to write *they was,* or *it depend,* or *the student have.* But there are four special sentence situations in which you are liable to go wrong in your singulars and plurals. We will illustrate all four.

The first is this:

> The *duties* of the guard *depends* on the type of post.

You know that the plural *duties* requires the plural form of the verb, *depend;* and you would naturally say *duties depend.* But the intervening modifier, *of the guard* (in which the noun is singular), lures you into a singular form of the verb, *depends.* To avoid this kind of error, put your one-word subject together with its verb and see if they match; the subject-verb *duties depends* is obviously a mismating; hence you must shift to *duties depend.*

The second sentence situation that causes singular-plural trouble is the *which-who-that* clause. This sentence will show the trouble:

> My roommate has two thoroughbred *calves* on his father's farm which *has* won several prizes at the Cattle Congress.

The difficulty is that, in this kind of sentence, the verb is often distant from the governing subject, so that the writer forgets what number he should use for the verb. To test this situation, simply put the subject, relative, and verb together—*calves* which *has*—and you will see at once that the verb should be *have.*

The third troublesome situation may be illustrated by this sentence:

> For me the most important thing in high school *were* the activities.

Here the plural word *activities* has coaxed a plural verb out of you: but it is the word before the verb, *thing,* which is the subject, and your sentence should read:

> For me the most important *thing* in high school *was* the activities.

If, however, you made *activities* the subject, then the sentence would read:

> For me *activities were* the most important thing in high school.

The fourth trouble spot occurs in the *there* type of sentence, where the subject follows the verb. Be cautious with *there;* it is merely a "function" word which gets the sentence started:

> In the new gym there *is* a swimming *pool*, a wrestling *room*, and a basketball *court*.

With a plural subject—*pool, room,* and *court*—a plural verb is needed:

> In the new gym there *are* a swimming *pool*, a wrestling *room*, and a basketball *court*.

Here is the same kind of error in another sentence:

> In tennis there *occurs* several *mistakes* that beginners should avoid.

Since the subject, *mistakes,* is plural, it should have a plural verb:

> In tennis there *occur* several *mistakes* that beginners should avoid.

REVISION PRACTICE

1. Each of his many friends have listed his positive traits.
2. While we remained, everything which belonged to the members of our group were there for the use of all of us.
3. The objectives of the program, of which I was fully informed, has made me less anxious to become a member.
4. The first and most important activity are the meetings.

Agr SV 1

5. The second and more important factor are the time trials.
6. Neither of these two things are true in college life.
7. The results I expect to obtain from my education is equal to the amount of effort expended.

ADDITIONAL REVISION PRACTICE

1. In our society today there exists many problems which can be solved only by social scientists.
2. In addition, the stimulus having predictive value rather than confirmatory value were preferred.
3. Learning the rudiments of music are not as difficult as learning to appreciate the different types of music.
4. Among the guests was an artist, a ball player, a writer, and a bank president.
5. Any one of these people are capable of doing the work.
6. I feel that the success may be attributed to the many hours of concentrated effort which is spent.
7. The law of averages protect you in a risky situation such as this.

2 Amb

Ambiguity

A word or passage is ambiguous when it can convey more than one meaning to the reader. For example,

How will he find his dog tomorrow?

To the writer this sentence probably seemed simple, direct, and clear. He did not notice that the word *how* can convey two meanings, *in what condition* or *by what means*. What happens in a sentence situation like this is that the writer chooses words which convey quite adequately the meaning in his mind, but he fails to perceive that these same words might have a different meaning for the reader. In this kind of sentence the ambiguity can be removed by substituting a term that has only a single meaning.

Now let us look at a different kind of ambiguous sentence:

What we believe profoundly influences our ability to listen fairly.

In this case the offending word *profoundly* has only one meaning, but the passage has two meanings, because we do not know whether the writer means *believe profoundly* or *profoundly influences*. The fault can be corrected by punctuation or by shifting the position of *profoundly*:

What we believe, profoundly influences. . . .
What we believe profoundly, influences. . . .
What we profoundly believe influences. . . .
What we believe influences profoundly. . . .

Here is still another kind of ambiguity that often crops up:

The course included the theory of procurement, property accounting, and requisitioning.

This kind is hard to see while one is writing, but it is easy to correct after one has spotted it:

The course included property accounting, requisitioning, and the theory of procurement.

or

The course included the theory of procurement, of property accounting, and of requisitioning.

There are many different ways in which a passage can be ambiguous, as you will see in the revision practice below. The important thing is to detach yourself from your script and read with a fresh eye, as if you were a reader seeing the page for the first time. This is hard to do, but one practice has been found helpful by many writers: put your next-to-the-last draft away for a few hours and then reread it before making the final copy.

REVISION PRACTICE

1. Mac's sister did not care for the pigeons.
2. Businessmen who are afraid to take risks frequently lose out to their competitors.
3. His job was to post changes in address, telephone numbers, and performance ratings.
4. The club needed more intelligent officers.
5. Every child awaits the time he can go to school with great excitement.
6. We were served fresh strawberry ice cream.
7. It was a long colorful parade with many decorated trucks, cars, and buses.
8. I like to be out where I can look for several miles in any direction without seeing more than three or four sets of buildings, trees, and animals.
9. Both political parties spent over three million dollars on radio and television.
10. I have football from 3:15 to 6:00; then the fun begins.

ADDITIONAL REVISION PRACTICE

1. The superintendent was observing the construction of the new school.
2. The three rooms, the library, the kitchen, and the study were smartly decorated.
3. He told me that I should have given more realistic details.

2 Amb

4. In my methods course we studied modern language teaching.
5. My roommate insisted on wearing a big plaid shirt.
6. The answers to the questions were all right.
7. Natural beauty is an inspiring sort of thing, like a Monarch butterfly landing on a peony or a rainbow after a refreshing rain.
8. The club will be open to members only from Monday to Thursday.
9. The examination of Dr. Smith was a long one.
10. They drove fifteen ton trucks.
11. The life of a movie star that the public sees does look glamorous.
12. This salve offers soothing relief for mild sunburn, poison oak and ivy.
13. In the new area there are a new school and a new hospital, built in 1955.
14. A baseball player must have good vision, coordination, and speed.

3 AP

Use of Active Instead of Passive Recommended

The passive voice is a convenient rhetorical device. It may be effectively used in at least four situations:

1. When the writer wishes to remain noncommittal:

    We suspected that the boat had been stolen.

Here the writer remains uncommitted as to who did the stealing.

    2. When a link with the preceding sentence is needed:

> This is one definition of literature. The other has already been discussed.

In this construction, *other* is placed close to its referent, *definition,* in order to aid the flow of ideas.

    3. When the doer is unimportant:

> The restaurant will be closed at nine.

    4. When the subject of a passive verb is the matter of greatest interest:

> A lovely red Persian rug was unrolled at our feet.

To use the passive effectively, you should be aware of two caveats. First, you should avoid "weak passives"—those which do not serve a purpose. These only lead to wordy constructions and false emphasis. Here are some examples:

> The biggest fish were caught by me and my brother.
> The next topic has already been discussed by my opponent.
> I was surprised to hear the view that was expressed by him.

These sentences are neither wrong nor unclear, but they would sound more straightforward if they were made active:

> My brother and I caught the biggest fish.
> My opponent has already discussed the next topic.
> I was surprised to hear the view he expressed.

Second, a sentence which begins in the active voice should not be shifted to the passive voice without good reason. Take this sentence for example:

> The speaker who does this believes that he is correct in the point which is being put across.

It would be more natural and direct to retain the active voice throughout:

3 Ap

The speaker who does this believes that he is correct in the point which he is trying to put across.

REVISION PRACTICE

Identify and correct active-passive shifts and "weak passives" in the following sentences:

1. In writing a biography the writer should be as objective as possible. A writer, however, cannot be completely objective, and care must be taken to avoid being taken over by one's own emotions and desires.
2. I was able to finish the first semester, but a lot of adjustments had to be made.
3. Once the lesson was studied and the paper written, plans were made for an exciting weekend.
4. When paddling a canoe across a rough lake, great care must be taken to avoid tipping.
5. At every performance of the symphony the music was enjoyed by me more.
6. Melody wanted to be warm enough on the ski trip. After she bundled up in her woolen skirt and ski sweater, her long scarf was found and put on.

ADDITIONAL REVISION PRACTICE

1. Archery can be participated in by any number of people.
2. While one is taking the trouble to solve one difficulty, another is noticed.
3. We sent them information, and ideas were exchanged in a spirit of warmth and cooperation.
4. The reviews on this new and popular play were disagreed with by the theater audiences.
5. He was handed his discharge by the president of the company.

6. She drove the car skillfully. After looking in both directions for cars, the gear was shifted, after which the clutch was released slowly and she was on her way.

4   Cap and LC

Capitalization
and Lower Case

The use of capital letters is a matter of either style or convention. In situations where the writer makes a stylistic choice he should be governed by the need for clarity and emphasis. For example, good writers often capitalize such words as *Church, State, Hope, Truth,* or *the Presidency* when they are referring to concepts important within the context of their writing (rather than simply using the word in a general sense). Writers may capitalize almost any noun (*Reproduction, Redundancy, Management, Labor*) to indicate that it is a key word. But be wary of these uses of capital letters.

Many of the conventions, such as capitalizing the first word of a sentence, the pronoun *I,* and pronouns referring to God, are so familiar as to need little explanation. In the following situations, however, freshman writers often forget to capitalize:

1. In titles, the first and last words and every word except prepositions, articles, and conjunctions are capitalized. Freshman writers often allow the shortness of a word to mislead them. Here are some typical errors in titles:

FAULTY: Explaining the signs of the Zodiac
CORRECT: Explaining the Signs of the Zodiac

FAULTY: The Attrition rate of College Freshmen
CORRECT: The Attrition Rate of College Freshmen
FAULTY: The Challenges of my new Position
CORRECT: The Challenges of My New Position

2. The words *east, west, south, north,* or combinations of these are capitalized when they refer to sections of the country, but they are not capitalized when they indicate directions. You would capitalize *west* in this sentence—

I learned about the old West from watching TV.

but not in this one—

You go west two blocks to reach the junction.

3. Months and days of the week are always capitalized:

In August the regular Tuesday meetings were suspended.

4. Official titles and names are capitalized. You would capitalize here—

Tonight I am going to hear the President of the United States.

but not here—

My brother is no longer president of the company.

Students sometimes forget that a relative's name is similar to a title.

Here are some examples:

FAULTY: The house has a big front porch where grand-
           mother and grandfather Jones sit every night
           after supper.

CORRECT: The house has a big front porch where Grand-
           mother and Grandfather Jones sit every night
           after supper.

FAULTY: On my tenth birthday I received a bicycle from
           uncle Harry and aunt Harriet.

CORRECT: On my tenth birthday I received a bicycle from
           Uncle Harry and Aunt Harriet.

5. Names of races, nationalities, and languages are always capitalized.

In our neighborhood there are Negroes and Italians living side by side.

We went to a restaurant that featured Chinese food.

The prisoners all spoke good English but with a German accent.

In correcting capital and lower case errors it may be helpful to remember that generic names (*zinnia, botany, freshman, president, horse*) are not capitalized, but that specific names frequently are (*Red Giant Zinnia, Botany 101, Freshman Week, President Lincoln, Man of War*). Note that specific course titles are capitalized, but subjects (*psychology, mathematics, history*) are not, except for terms like *American history* and *French literature.*

REVISION PRACTICE

1. When I found they were open, I immediately registered for history 220 and biology 106.

2. Their debate topic year after year was the same: "Should Red China be admitted to the United Nations."

3. He first came to the middle west early in his career and later settled permanently in the east.

4. In Autumn life was dull indeed but Spring was a different matter.

5. She is a mythical maiden from the area now called the holy land.

6. He was hard pressed to remember which were Greek Gods and which were Roman Gods.

7. After labor day he no longer felt the urge to go north on fishing trips.

8. His Uncle Fred had given him the watch on the tuesday before he had graduated from Mt. Morris high school.

9. Most College Graduates have little difficulty in finding jobs in this field.

10. Nobody read professor Harold's book unless forced to read it.

ADDITIONAL REVISION PRACTICE

1. The course in American Literature presented no problem but he had his usual difficulty with Psychology and Mathematics.
2. In the living room there are three chairs, an Admiral Television Set, and a Duncan Phyfe Table.
3. Debating and Dramatics were extracurricular activities he enjoyed.
4. He gave his theme the title, "That Government is best Which Governs Least."
5. Excellence of football in the East is being challenged by the southwest.
6. Her Zinnias were not the Red Giant variety.
7. Bob Jones had been elected President of the Company.
8. Look up these words in the *American College Dictionary* as well as *Webster's new collegiate.*

5 CC

Comma or Commas
for Clearness

A comma will often help your reader avoid misreading your paper. It will show him which words are grouped together and will indicate where he should pause. Sometimes, for lack of a comma, the reader will misread momentarily, and then discover his mistake and have to read the passage again. Here is an example of such momentary confusion caused by the lack of a comma:

About two weeks after I started to work.

In this sentence the reader bumps into the period with a shock. Then he rereads and discovers that *after* is grouped with *about two weeks* and that there is a slight pause following *after*. A comma would have obviated all this trouble:

About two weeks after, I started to work.

You will often locate the places where a comma is needed for immediate clear reading if you will read your paper aloud and pause or change the pitch of your voice only where you have a comma. If the sentence makes sense when read this way, the punctuation is probably correct.

REVISION PRACTICE

1. After we had finished the books were returned to the library.
2. The room itself is small and empty boxes take up most of its area.
3. My home town is located near the state line and Freeborn County, Minnesota, to the north is a dry county.
4. When the speaker is moving people are not likely to leave the lecture hall early.
5. In this hotel James as you should have already known courtesy is the byword.

ADDITIONAL REVISION PRACTICE

1. Since the quantity is an unknown science has adopted the symbol X.
2. In much the same way a man may decide he should delay marriage until later in life.

3. The house he liked best was near to the shopping center and Clear Lake to the south was only minutes away.
4. As a young man dating was my favorite form of relaxation.
5. About three miles beyond the service station could be used to identify the turn for the club.

6 CF

Comma Fault
or Comma Splice

A comma fault results from the use of a comma at the end of a word-group where you should have a period, question mark, or semicolon. Here is an example:

> This may require as many as ten interviews, however, the counselee should leave each interview with a feeling of satisfaction.

This sentence, as you can readily see, is confusing to the reader the first time through. It should have been punctuated:

> This may require as many as ten interviews; however, the counselee should leave each interview with a feeling of satisfaction.

or

> This may require as many as ten interviews. However, the counselee should leave each interview with a feeling of satisfaction.

With this punctuation the writer has guided the reader to his meaning.

The next example also has a comma fault, but this one requires a different revision:

> Dick was hurt, he won't be able to play.

The best way to revise it is to show the relationship between the two parts (*see* Section 36 Sub) by putting one in a dependent clause:

> As Dick was hurt, he was unable to play.

There is still another way to revise comma faults, as we show in the next example:

> College has finally begun, about two weeks have gone by.

In cases like this you can connect the two clauses with a simple conjunction like *and, but, for, or, yet, nor:*

> College has finally begun, and about two weeks have gone by.

The comma fault is easy to avoid. You have merely to listen to yourself read aloud. At the ends of certain word-groups your voice will drop to a lower pitch and will dwindle away to momentary silence: a pause. This voice behavior is a vocal completion signal that, in writing, you indicate by a period, a semicolon, or a comma plus a simple conjunction, as in the examples above. Vocal behavior is not an infallible test, but if you will keep in mind two exceptions, it will prove serviceable to you as a convenient rule of thumb.

The first exception is illustrated in these examples:

> Mack lent his car to his brother, who promised to return it in two hours.
> The fire swept through the old barn, which was soon a smoking heap of embers.

In each of these sentences your voice may drop and dwindle away to a slight pause, yet the comma is the proper punctuation. The key to this situation is the use of *who* (*whose, whom*) and *which*. These words never begin a sentence unless it is a question.

The second exception is a variation of the preceding one:

> Mack lent his brother the car, a battered old Ford.
>
> The fire swept through the barn, a ramshackle structure with boards dry as dust.

At the comma position in these sentences your voice may drop, for at this point you have completed *a* sentence, but not *this particular sentence*. What follows the comma in each sentence is merely a modifier (called an appositive), not a sentence with a subject and a verb. Hence the comma is correctly used.

REVISION PRACTICE

1. We use many more Anglo-Saxon words than words of Latin or Greek, you will notice they are short, useful, necessary words.
2. There is one instance where I disagree with Ruskin, this is where he suggests going back to the life of the "rustic."
3. Around twelve o'clock I went down to lunch, I have long since forgotten what we had.
4. It is difficult to get interested in the subject, there are also times when one has no time for it.
5. The plane did not prove successful, however, the Germans perfected what the Italians had invented.
6. After the ball game they attended the dance, they had forgotten to tell their parents about these plans.
7. I miss the sun and the sea, I haven't adjusted to being back.

ADDITIONAL REVISION PRACTICE

1. The whistle blew, and I tried a reverse, my reverse failed and I ended up flat on my back.

2. It had rained and frozen the night before, therefore, the trip we had planned seemed risky.

3. Federal Aviation Administration regulations require that the airlines demonstrate their ability to evacuate a fully loaded jetliner in less than 120 seconds, finding up to 250 volunteers to leap from an airplane into the dark has become a problem.

4. I soon began to learn where certain buildings were and where my classes would be held, knowing these things gave me a bit more confidence.

5. It is uncommon when players with the immaturity of Romeo and Juliet are judged to have the maturity deemed necessary to essay the roles, these players were not so judged.

6. She had finished the sketch the day before, she was therefore clearly ready to enter it in the contest.

7 C

Uses of the Comma

More than 80 percent of the punctuation marks used inside the sentence are commas. Hence it is important that you master the uses of the comma if you are to guide your reader accurately along the trail of your thoughts. The dictum that a writer uses commas to represent slight pauses is useful, but you will do better to learn to apply the simple set of rules given below. These rules, together with those explained under Sections 5 CC and 22 PAC, will enable you to handle most cases of comma-punctuation that will come up in your theme writing. Commas are used:

1. To set apart interrupters, such as words, phrases, or clauses that interrupt the flow of thought.

> He is, by and large, the wealthiest farmer in the country.
>
> Jim, while a congenial fellow, is not qualified for the position.

2. To separate long introductory phrases and clauses from the rest of the sentence.

> Since no athletic scholarships are offered to any students of any university in the conference, the teams deserve to be called amateur.
>
> At a point near the heart of the business district in an ugly little town, the robbery was committed in broad daylight.

3. To separate the main clause from a long clause or phrase that follows it, if the two are separated by a pause or break.

> The outcome of the voting had been predicted, although student polling had not been done with any noticeable degree of enthusiasm.
>
> It had been done before, the problem remaining the same each time anyone had tackled it.

4. To separate two long independent or main clauses joined by a conjunction: *and, but, or, for, yet, nor.*

> All the city west of the river lay under a blanket of fog, and this made it impossible for traffic to move in any direction the rest of the night.
>
> I can't account for my actions on this particular day, but I know that I was not in the vicinity of the parking lot near the river.

5. To separate items in a series.

> He needed new tires, a new muffler, and a new battery for his car.
>
> We looked in the closet, under the bed, and in all the drawers.

When dealing with adjectives in a series, you must distinguish two kinds of situations. If each adjective relates directly to the following noun, then apply the rule above.

> a square, two-story, colonial house

But if each adjective seems to modify everything which follows, do not use commas.

> a big old friendly house

These two situations are sometimes hard to distinguish. If you are in doubt, use the commas.

6. To set off *yes* and *no*.

Yes, I can come early if you wish.

No, she has not been informed of the vacancy.

7. To set off words of address.

Bring me the book, Harriet.

I told you, Jim, that you might not be able to finish on time.

8. To separate the names of geographical locations where one location is included within the boundaries of the other.

They have a home in Claremont, California.

The license plate indicated that he was from Cook County, Illinois.

9. To separate the date of the month from the year. (A comma between the month and the year is optional.)

The family flew from Paris on May 19, 1968, for New York.

He became a citizen in June, [or no comma] 1955.

10. To set off titles and degrees from preceding names.

He was listed in the directory as Jackson Holbrook, Jr.

The name of the new staff member appeared on the program as Willard Ramsey, Ph.D.

REVISION PRACTICE

Punctuate the following sentences:

1. It was not necessary in my opinion to advise her to take such a heavy schedule.

2. The main speaker while his talk was pertinent and informative encroached on the time that had been set aside for entertainment.

3. I mentioned it earlier Jerry and had not expected you to bring the subject up again.

4. In a sheltered spot at a place near the road's end the body was hidden.

5. Joe had made the announcement earlier in the evening in front of all his friends but it was embarrassing to him to realize that practically no one wished him well.

6. It however was not a matter of concern to one but to all.

7. I explained to you John and to you Larry that one more revision was absolutely necessary.

8. On November 11 1918 the first war ended and everyone thought it was the war to end all wars.

9. The album included such titles as "Secret Love" "The Last Time I Saw Paris" "Lullaby of Broadway" and "It Might As Well Be Spring."

10. After rumors of foul play had been circulated he felt in this instance it was necessary to give out all the information he had.

ADDITIONAL REVISION PRACTICE

1. She had acquired at the auction a set of unmatched dishes a gone-with-the-wind lamp a tea cart some old books and some very bad paintings.

2. They had moved from Little Rock Arkansas to Galveston Texas all in a matter of weeks.

3. She had a list which included such items as clothespins detergent clothes basket clothesline and dye.

4. Yes it is quite likely that he will be elected to the student senate.

5. After dinner had been served and all the guests had been seated in the living room the readings began.

6. Franklin Townsend Jr. spent a life of misery in the town where he was born.

7. If however she had made it very clear in the beginning to all who had shown an interest in the project that they would be substantially rewarded she would not be in the difficulty that she is in now.

8. He had called for loyalty sacrifice and the good old college spirit and yes he had convinced them that these things were necessary for a winning team.

8   Cl

Cliché

A cliché (pronounced *clee shay'*) is an expression which has been overused so much that it no longer has any vividness or punch. A good example is *as strong as an ox*. Long ago, when the ox bending to the plow in the fields was a common sight, this expression must have conveyed a vivid impression of power. But today, used and heard by people who have never seen an ox, it is flat and stale. Our language is filled with clichés. They are handy counters in the give-and-take of ordinary conversation but have little place in careful, written prose. Here are a few, which will doubtless recall many others to your mind: *as old as the hills; stood like a sentinel; as bald as a billiard ball; last but not least; sadder but wiser; age before beauty; as brown as a berry.*

The remedy for a cliché is to substitute fresh and vigorous words that will vividly indicate what you wish to convey. Instead of *as brown as a berry,* why not say *as brown as the last oak leaf,* or *as brown as the hills near Point Lobos,* or *as brown as a Hereford,* or *as brown as cinnamon*?

Another way to replace a cliché is to use a simple and direct expression. For *pull down a victory* use *win.* For *rears its ugly head* use *appears* or *presents itself.*

REVISION PRACTICE

In revising the following sentences, try to strengthen them by using fresh, imaginative expressions or simple direct terms, instead of the clichés.

1. Where else could you find out that you are not entering a cold, cruel world?
2. To many students the lagoon is a place where one can get away from the hustle and bustle of campus activities.
3. People trying to save money will do without this or that and never go overboard on anything.
4. His idea does not hold too much ground in the twentieth century.

ADDITIONAL REVISION PRACTICE

1. Life is simply not that cut-and-dried.
2. Chances are that truer words have never been spoken.
3. The lake is surrounded by trees that sway to and fro in breezes cooled by the waters.
4. Being around so many extroverts brought her out of her shell.
5. Getting money out of my roommate was like pulling teeth.

9   Col

Use of Colon
Recommended

The uses of the colon are easy to understand. First, like the dash (*see* Section 12 Dash) it is used to signal to the reader that a series will follow:

> During our January sale we are featuring the following items: boys' T-shirts, shorts, and socks; men's slacks and underwear; and ladies' raincoats, hats, and overshoes.

It is often regarded as poor practice to separate the main elements of a sentence—subject, verb, and object—by a colon:

> We saw: battleships, cruisers, carriers, and destroyers.

You will notice immediately that the colon separating the verb from the object is unnecessary:

> We saw battleships, cruisers, carriers, and destroyers.

However, in a long, more fully developed sentence this use of the colon can be helpful:

> After the first attack had subsided and we had had several hours to rest, the chief gunner reported: three battleships, all accompanied by escort vessels; thirteen troop carriers; an undetermined number of aircraft carriers, all of which seemed to have planes at the ready; innumerable PT boats.

Second, the colon indicates that the next clause to come will clarify, expand, or illustrate the idea just mentioned. Here are some examples:

> There is no time for love: we are doomed.
>
> To deny this is to deny the basic premise of the Declaration of Independence: "All men are created equal."
>
> I had seen her before so I knew what to expect: she was all skin and bones.

The colon serves the same purpose in introducing a long quotation:

> He made these interesting comments concerning the character of Iago in *Othello*:
>
> Iago seems to be the enemy of love. In fact, he seems to be the enemy of all emotion. He spins ruthlessly a web of jealousy in the mind of Othello.

REVISION PRACTICE

Provide a colon where appropriate, or correct the misuse of the colon in the following sentences. Rewrite when necessary.

1. For actors of today I know some other excellent advice Hamlet's speech to the players.

2. This is the reason we are returning to humanism because the world situation requires it, not because man is improving his character.

3. The question is this. Should we agree to settling problems such as these in such a haphazard and selfish manner?

4. He had one idea in his head to get out of the room at once and in a socially acceptable manner.

5. One is asked for the following types of themes. They are definition, description, comparison, and clarification.

6. For your next assignment please read an article from one of the following magazines *Harper's, The Atlantic, Scientific American,* or *Holiday.*

ADDITIONAL REVISION PRACTICE

1. He had three diverse hobbies stamp collecting, swimming, and drag racing.
2. He had a very good reason for not coming to the party he was not invited.
3. The main question is this. Should the examination be uniform throughout the United States or should it be so constructed as to fit the problems peculiar to each state?
4. Creon thus summarized the theme of Oedipus the King "Do not seek to be master in everything, for the things you mastered did not follow you throughout your life."
5. There was also the social purpose to meet students from different areas.

10   Colloq

Colloquial

A colloquial expression is one that is popular and acceptable in conversation and informal writing but is not used in formal writing. A few familiar examples of colloquialisms are *kind of* (rather), *fizzle out, goner, mighty* (as in *mighty pleased*), *awful* (as in *awful glad*), *go back on,* and *real* (as in *real good*). Since a formal style is required for most college themes and papers, you should avoid colloquial terms in such writing.

Your desk dictionary will help you with usage problems like this one. Three of the desk dictionaries usually recommended for college students label these terms *Colloq.* or *Slang*. A fourth recommended dictionary, *Webster's Seventh New Collegiate,* does not use either of these labels, apparently in the belief that the status of such terms cannot be accurately ascertained. There are occasionally differences among the four dictionaries in their choice of usage labels; for instance, a word labeled *Colloq.* in one may be classified as *Slang* in another. In either case, the word indicated should not be used in formal writing.

You must also remember that a usage label is applied only to a specific use of a word, not to all of its meanings. The word "comeback," for example, has three definitions in one desk dictionary. One of these is standard usage, the second is slang, and the third is colloquial.

11 Conf

Confused Sentence

The term *confused sentence* is a coverall label that your instructor will use for sentences that are muddled and disorderly, with their parts inserted helter-skelter. There is no single way to go about revising such sentences (*see* Sections 23 Paral, 36 Sub, and 20 OBS). Often, however, one or more of these procedures will be of help to you: (1) take a deep breath and make a new beginning; (2) rearrange the parts; (3) use two or three sentences instead of one. Let us examine this confused sentence:

> In a rural school there is usually no basement, poorly kept floors, outmoded equipment, poor heating, and the floors are extremely drafty in winter.

Now notice what has been done to make this readable:

> Rural schools are in poor condition. They usually have no basement, their equipment is outmoded, and the heating is inadequate. Their floors are poorly kept and are extremely drafty in winter.

Here is another:

> In shorthand you learn the grammar parts of letters about paragraphs, commas, colons, and other punctuation.

This can be simply and directly stated:

> In shorthand you learn two aspects of letter writing—paragraphing and punctuation.

REVISION PRACTICE

1. He could be found in the morning at two o'clock in a little gas station on weekends.
2. My sentimental thoughts are of my home town. It is a very small town out in the middle of the country and all of the farms.
3. In a study of this type emotions would eventually enter, no longer making the study objective.
4. The divisions of freshman English coincide so that some of each division is done at the same time, such as essay reading and theme writing.
5. The marshals usually dressed in red fighting to keep the huge crowds back that line the fairways, all waiting for a chance to see their hero or the tournament leader.

ADDITIONAL REVISION PRACTICE

1. Intramural sports play a very important part in every college curriculum through building sportsmanship and also teaching and learning the games to individuals for the enjoyment of the sport.
2. The amazing thing about some of these advertisements, that people just hate and could throw a shoe through the television screen or toss the radio out of the window, do succeed in selling their product.
3. My mother informed me I was an unhappy girl according to Miss Smith and asked that my mother if at all possible have me quit my job.
4. The third, and an important, step is selecting the right kind of tools and equipment. Suggestions will probably be found in seed catalogues, with regard to the types of plants, and an experienced gardener is always free with advice. The gardener must keep in mind the size of the garden. . . .
5. In high school, extracurricular activities were both during class hours and after school. This idea of class and after-school activities gave the student a full schedule and showed the student how to fulfill it in many ways as becoming well acquainted with the activity, becoming active in the activity and give the students the full cooperation to the activity when having meetings or group activities which fall into the extracurricular activity.

12 Dash

Use of Dash Recommended

Among the functions of the dash, two are especially important. The first is the use of a dash, or dashes, to set off a series as a unit from the rest of the sentence. Here are three examples, with the unit-series in the beginning, the middle, and the end positions:

> Sweaters, shoes, old magazines, used scratch paper, a catcher's mitt—these had been tossed under his bed in a tangled confusion.
>
> We have thus seen that certain qualities of character—a sense of humor, enduring patience, and a genuine fondness for young people—are necessary for success as a classroom teacher.
>
> In addition to sports he had numerous other strong interests—painting animals, collecting unusual rocks, cooking on the beach, and raising hamsters.

In these examples we see that in each case the unit-series is preceded or followed by a coverall word—*these, qualities,* and *interests.*

A second useful function of the dash is to set off an interrupting statement or term.

> My father had always insisted—and he had an insistence of iron—that one should always examine the second side of any argument.
>
> We hiked down the coomb—known to you Americans as a gully —to search for the groundhog's den.

REVISION PRACTICE

1. These three processes: leveling, sharpening, and assimilation tend to twist facts into rumors.
2. The trick knee, the misshapen nose, the weak ankles, all these will remain with the man throughout his life.
3. He had just one idea in mind to get the money the quickest way possible.
4. He was told though he did not listen that further difficulties with administrative officials were to be avoided.
5. Some men Howard K. Smith, for example were born news commentators.

ADDITIONAL REVISION PRACTICE

1. The intense struggle for power, it had now been going on for a year, was about to be resolved.
2. Several football players, Jones, Roberts, and Hanson, were about to be dropped from school.
3. It was the night of September 5 when the boy, I still can not remember his name, finally decided the time was right.
4. The sunrise, the soft wind from the east, the smell of bacon in the air this combination, appealing to the senses, made getting up a necessity.
5. The little plot was filled with common garden flowers pansies, marigolds, nasturtiums, and larkspur.
6. One thing was known for certainty that Cox was not going to be around very long.
7. The real reason why the program failed all his protests to the contrary was that the chairman shirked his responsibility.

13 Frag

Fragment

A fragment is a part of a sentence which has been written as a whole sentence, that is, with a capital letter and a period. Here are three examples:

> Everyone was tense with excitement waiting for the sound of the final gun. Which did not go off until fifteen seconds after the clock indicated the game was over.
>
> He was covered with mud from head to foot. Having practiced on a muddy field for two hours.
>
> The criminal, the one who lives in the darkness of night's concealment, many times forced to the edge of insanity by fear.

Usually a fragment belongs with the preceding sentence and should be preceded by a comma, thus:

> Everyone was tense with excitement waiting for the sound of the final gun, which did not go off until fifteen seconds after the clock indicated the game was over.
>
> He was covered with mud from head to foot, having practiced on a muddy field for two hours.

Sometimes fragments can be made complete sentences by the inclusion of a subject and a verb, as in this case:

> The criminal, the one who lives in the darkness of night's concealment, is many times forced to the edge of insanity by fear.

Fragments are sometimes deliberately employed for a specific purpose by professional writers, but those who are learning to write are advised to avoid their use.

REVISION PRACTICE

1. The suggestions made so far add up to one thing. That this investigation is not necessary.
2. One can begin his task in another way. By using all his spare time in organizing his material.
3. There is a rapidly growing field concerned with the study of mass media. And with the impact this media has on society.
4. One night during the summer we had a very heavy rain. Which caused the creek to rise to a dangerous level.
5. She had poise and intelligence. At the same time giving the effect of being unusually pleasant.
6. For some distance the wide road is hemmed in by trees on both sides. Thus giving the driver a feeling of security.
7. The reading of textbooks which are not used at this college. But which are useful as guides in course study and reference reading.

ADDITIONAL REVISION PRACTICE

1. One might also think that this was an important issue. For the simple reason that it headed the agenda.
2. The designed program should provide a unique training program for those in attendance. This with an appropriate amount of time for social activities.
3. The following day he felt even worse. Which could have been a factor in his miserable performance.

4. The boys, who decided they were no longer interested. They were the very ones who had insisted on holding the meeting in the first place.
5. We are asking you to come earlier than the previously announced time. With everything that you were requested to bring.

14   Imp

Imprecision

The term *imprecision* is applied to passages in which the writer does not say precisely what he means, or in which he says something that is evidently impossible. Here is an example of imprecision:

> Many people have physical disabilities that they do not know about. One of the most important of these is vision.

In this passage the writer has actually said that *vision* is an important physical disability, but vision is not a disability at all. What he seems to have meant is:

> Many people have physical disabilities that they do not know about. One of the most important of these is defective vision.

Here is a second example:

> The importance of good physical condition, developed by a rigorous training program, is the most basic of all football fundamentals.

In this sentence the writer has said that *importance* is a basic football fundamental; but what he probably meant was:

> Good physical condition, developed by a rigorous training program, is the most basic of all football fundamentals.

In the careful writing that you are expected to do in college, it is imperative that you write with precision.

REVISION PRACTICE

1. I thought eye contact and speed were among the significant weaknesses in the delivery of the speeches.
2. These trees are the tallest things in the world.
3. Another of the important principles of a good basketball player is speed.
4. This is one conception of Communism. Another, particularly used by intellectuals, is a person who believes that the theories taught by Marx are the answer to our economic ills.
5. The most important factor leading to my decision was because of the principal of my school.

ADDITIONAL REVISION PRACTICE

1. The basic idea of a good baseball pitcher is to know his arm.
2. A good example of the core program is illustrated by the high schools of Los Angeles.
3. Reading is a necessary part of life, if one is to keep abreast of current events, such as general information pertaining to worldwide and local interests.
4. They can join organizations that they would like to further their ability in.
5. Subject matter, organization, eye contact, and the

speaking voice are factors to be developed in public speaking. Constant practice is the only way to overcome each of these points.

15 M

Meaning Is Not Clear

The symbol *M* indicates that you have not made your meaning clear. It is doubtless clear to *you,* but you have either not chosen the right words, or have not arranged them in the right way to make your meaning clear to the reader. (Sections 14 Imp and 15 M overlap. In 14 the reader can usually guess the poorly stated meaning. In 15 the meaning is blurred. For some errors, either symbol, Imp or M, will do.)

To revise the fault you must rewrite the offending passage. Sometimes it helps to ask yourself, "Now, in plain and simple English, just what is it that I want to say?" and then answer your own question and write down the answer. Another way to get a clear statement is to read a doubtful passage to a friend. If he understands what you meant and can explain the meaning to you, then you can be reasonably certain that the passage is at least clear.

Here is an example of a sentence that would be labeled with the marginal symbol *M :*

> [The writer is describing the reference room of a college library] In the center is a large card file surrounded by tables covering the rest of the floor and bookshelves along the wall.

The meanings here splatter out in all directions, as you can see if you study the possibilities. Yet just a little more care in phrasing could have produced a clear sentence like this:

In the center of the floor is a large card file surrounded by tables, and by bookshelves along the walls.

REVISION PRACTICE

1. West Branch has a breakdown of about five religions.
2. On both sides of the hall is located an office.
3. Another law frequently broken is exceeding the speed limit. The law was made because of the condition of many of our highways, compared to our modern high-speed automobiles.
4. I have discovered that while giving a speech, there are several very important devices that help give clarity and expression to the speaker's subject.
5. There are also seventeen gasoline and service stations in the city, three of which include either a motel or cabin court in addition to the Clark Hotel.

ADDITIONAL REVISION PRACTICE

1. When a person puts his shoes on he reveals his character to others, who are the judges. Shiny shoes with clean shoestrings always give a neat appearance. Good character is the result.
2. One should stand comfortably erect and emphasize his points with gestures—with either a movement of the hand or perhaps a nod of the head. These devices give the speaker poise and also enable him to speak in a clear distinct voice, adding clarity and emphasis to his talk.
3. The Webster City squad suffered only three major injuries which was reflected in their brilliant season record.
4. Duty may be something that we must pay in legal tender for the privilege of doing some specific thing.

5. Going through high school in this day and age is no more than the elementary schools were years ago.
6. College is not at all as I expected it to be. Of course staying at home does not help the situation any.

16 Mis

Misplaced Modifier

A misplaced modifier is a word or word-group which has been so placed that the reader is not sure what it goes with. Of course, he can usually find out by reexamining the sentence, but he should not be required to waste his time in this way. Besides, if the reader is perplexed—or amused—his thoughts are diverted from what the writer has to say. A sentence with a misplaced modifier can usually be satisfactorily revised by placing the modifier in a different position, though occasionally a rearrangement of other sentence parts becomes necessary. In the example which follows, the writer says something that he probably does not mean:

> I spent the afternoon talking about books I had read with the librarian.

A simple change in the position of the modifier will make the sentence clear:

> I spent the afternoon talking with the librarian about books I had read.

The following misplaced modifier may cause the reader to smile with amusement and forget what the writer is saying:

He has a blue satin ribbon around his neck which is tied in a bow.

To correct this, we need only to change the position of one little word-group:

Around his neck he has a blue satin ribbon, which is tied in a bow.

If you take pains with the arrangement of the parts of the sentence you can avoid the embarrassment of misplaced modifiers.

REVISION PRACTICE

1. They stood watching the parade in the back yard.
2. After that, Harvey came to the band practices that I called with his horn.
3. We saw the ambulance driver, Mr. Jones, who was picking up the injured, covered with bruises and congealed blood.
4. There is also a bar located in the heart of town which is filled to capacity every evening.
5. He had an old radio that he later traded for a new television set which he loved to tinker with.
6. Every sixteen-year-old looks forward to the time he takes his driving test with great excitement.

ADDITIONAL REVISION PRACTICE

1. We have houses for middle-income people with built-in kitchens.
2. This movie portrays life as it is in a well-written plot.
3. I noticed a small window in the tunnel which was broken.

4. I have this opinion because I have seen several youths while under the influence of alcoholic beverages.

5. You will notice a small circular disk protruding from the watchcase, which is commonly referred to as the stem.

17 Mis OM

Misrelated
Opening Modifier

A modifier is often placed at the beginning of the sentence. The following sentences will illustrate:

> Holding the guide rope carefully, Pierre felt his way down the narrow rocky trail.
>
> Before replacing the faucet, you should make sure that the water is turned off.
>
> Tired and dusty, the hikers stumbled into the inviting shade of a large wayside elm.
>
> To unlock the door, one should turn the key to the left.

When a reader meets such opening modifiers, he tends to connect them with the nearest word possible, which is usually the subject of the sentence. Hence it is desirable, though not always imperative, that the opening modifier modify the subject. But what is most important is that the meaning be clear and that the opening be free from unintended humor. When Nixon was Vice President, he made a slight blooper in the use of an opening modifier; listeners to and readers of his speech had reason to smile:

As recognized leader of the Truman wing of the Democratic Party, I advise Mr. Stevenson that if he has a better program to offer. . . .

Opening modifiers like the following are not really wrong or unclear so much as clumsy or amateurish:

In driving a heavy nail, the hammer should be held near the end of the handle.

Seeing the juicy steak sizzling over the charcoal, his mouth watered.

To catch black bass in these waters, live minnows should be used.

But a careful writer would present them in this manner:

In driving a nail, you [or one] should hold the hammer near the end of the handle.

When he saw the juicy steak sizzling over the charcoal, his mouth watered.

To catch black bass in these waters, one should use live minnows.

No hard and fast rules can be given for opening modifiers except that they should be clearly related to the rest of the sentence and carefully constructed. It is best to relate them to the subject of the sentence.

REVISION PRACTICE

1. By reading the questions carefully and thinking out the correct answers, a lot of things in the examination will be remembered for a long time.
2. When designing a jet aircraft of any type, there are many things that the engineers must take into consideration.
3. To allow plenty of room for the roots of the shrub, a deep hole should be dug by the planter.

4. While registering, my class cards got lost.

5. Speaking in this manner, the warmth and meaning of the word is lost.

6. In stating these three points, the uselessness of God has been proven by Huxley.

ADDITIONAL REVISION PRACTICE

1. Having improved my writing, these punctuation exercises have been to my advantage.

2. Looking around, a few cars were parked when my husband and I first arrived.

3. Shortly after arrival, these same beaches are used for canoe orientation and lifesaving techniques.

4. As a student, reading is the most important basic skill on which my success depends.

5. By working toward this end, peace seems much more realistic.

18 MM

Mixed Metaphor

A metaphor is a statement of likeness between unlike things. A few examples will illustrate:

> Instead of studying a history assignment from beginning to end, he merely frogs through it. (A student and a frog are unlike things, but they have a link of likeness in the way the student leaps through his assignment.)

In the opening lecture the professor spun his wheels for about ten minutes before he began to move ahead.

Once on the crowded dance floor Lou lost his rudder and seemed to collide with every nearby craft.

Metaphors enable your reader to associate concrete images with the ideas you are presenting. These images, if fresh and pertinent, help him to understand and remember. But if you forget to image for yourself what you are asking your reader to image, you may employ an incongruous metaphor or may mix up two metaphors. In either case the effect can be confusing or unintentionally humorous. Here are two examples:

The shafts of the sun gave a shot in the arm to the gloomy afternoon.

He climbed the ladder of success on the shoulders of his friends.

These are incongruous and mixed respectively.

REVISION PRACTICE

In revising the following sentences, reconstruct the metaphors or eliminate them, especially the clichés.

1. The politician must straddle the fence with the skill of a tightrope walker, while staying on the right side of the fence that keeps his conduct and character above reproach.

2. The snow, like a giant mirror, reflected the moonlight, lending a helping hand to the weary travelers.

3. The Secretary of State seized the bull by the horns and flew to Uruguay.

4. Man has made great strides, but there are still many stones left unturned.

5. Bacon says that some books are to be tasted, some swallowed and others digested. English composition is a stepping stone to help us become digesters.

ADDITIONAL REVISION PRACTICE

1. The college merry-go-round has always seemed so glamorous that everyone is ready to jump on without weighing it in the balance.
2. The windows look as if they are merely transparent sections in a sea of steel.
3. One can feel the caress of the winds as they reel across the roving lands.
4. Whitwood and I jumped into my convertible and set sail for faraway places.
5. If you are going to wade through freshman composition, you must really get your teeth in it.

19 NS

Nonstandard

Nonstandard English is English which does not conform to the grammar, word choice, or idiom of its educated native users. It is the language of the uneducated and—even though it may be clear, vigorous, and colorful—has no place in college themes. Here are three examples of nonstandard written English:

1. (GRAMMAR)   The letter he had wrote was lost
2. (WORD CHOICE)   He busted his arm
3. (IDIOM) All to once, the dam burst.

REVISION PRACTICE

Rewrite, removing the nonstandard expressions.
1. Most of the freshmen girls was at the rally.
2. A grey nondescript cat was setting quietly on the doorstep.
3. The next contestant drug his bawling calf to the arena.
4. Our carpenter refused to use them split boards.
5. Where was you when I phoned?

ADDITIONAL REVISION PRACTICE

Rewrite, removing the nonstandard expressions.
1. Jean was determined to attend the conference irregardless of her sprained ankle.
2. In the tidal wave that followed the quake many persons were drownded.
3. The chart was laying right in front of him.
4. George and him carried the injured man on a stretcher to the waiting ambulance.
5. Our center played very good during the last quarter.

20 OBS

Overburdened Sentence

An overburdened sentence is one that is too long and too heavy. Such a sentence is hard to read and should be broken up into two or more sentences. The inexperienced writer tends to continue a sentence on and on, fighting his way through a thicket of clauses and phrases, not knowing when or how to stop. This may be all right for the first draft, where one is getting his ideas down on paper regardless of form and style; but in the first revision one should spot all such marathon sentences and segment them into shorter sentences.

Here is an example of a long-winded sentence that puts strain on the reader:

> The plot includes Madame de Renal, who is married to a boring husband who thinks only of his personal prestige in society, Mademoiselle de la Mole, who is bored, rejects her boring suitors, and when she does meet an ambitious and exciting man, she remains restrained because of her noble birth and high position in society, Julien Sorel, the hero and the third son of a poor Verrières water-mill worker, who in spite of his humble birth, becomes involved with both Madame de Renal and Mademoiselle de la Mole.

Toward the end we also notice that the writer has lost control and has produced two ambiguities. Let us try to improve this sentence without altering the sense.

> The plot includes a varied array of characters. Madame de Renal is married to a boring husband who thinks only of his personal prestige in society. Mademoiselle de la Mole, who is bored, rejects her boring suitors; and when she does meet an ambitious and exciting man, she remains restrained because of her noble birth and high position in society. The hero is Julien Sorel, the son of a poor Verrières water-mill worker. In spite of Julien's humble birth, he becomes involved with both Madame de Renal and Mademoiselle de la Mole.

This revision is now clear and readable. For further help in rebuilding overburdened sentences, *see* Sections 23 Paral, 36 Sub, and 24 Per.

REVISION PRACTICE

> 1. A person in high school may be exceptionally large and
>   go out for football not because he wants to but because

if he does not he may think that people will talk about him and call him "sissy" or maybe a person is very tall and goes out for basketball not because he likes the game itself but because he is afraid of what people will say if he doesn't.

2. Another of my most powerful motives for choosing Mrs. Finney "The Best Teacher I Ever Had" was because of the understanding and ability to help someone out of a low spot which they would invariably suffer during the course of a semester, and instill instead a highly confident knowledge that sooner or later you would pull out of your more or less mental slump and fight your way through the mental blocks that now stand.

3. Professors should be very careful about grading papers where there is no set dividing line to differentiate between grades, for a paper that is poorly graded could result in the further downfall of the student, and a paper that is graded to the best of the professor's ability could lead to better work on his part, and though of course this does not hold true for every student, in general this is true.

4. His future was assured and he knew it, so he supported every wild plan anyone could think of, and he never worried about a thing like the rest of us who were never certain where our next meal was coming from and often didn't particularly care, although we were sensible enough to know that we had to meet at least simple needs to carry on.

5. If Tom had only known what to do and the correct time to do it, he would have been able to stay in school but he was a poor planner and a poor organizer and his failure could have been predicted, for these faults are predictors of failure in academic work and for any other kind of work for that matter.

ADDITIONAL REVISION PRACTICE

1. Stevenson portrays his main character as a person who started taking the easy way out of working by stealing and shows that even though Markheim was nervous and frightened at what he did he kept on committing crimes until he finally committed murder.

2. Finally as I began to get all my classrooms located and bought most of the books I needed I thought back on how much more confusing this week at college would have been had it not been for the professors and members of the university staff who took time off and went to extra trouble to help the incoming students feel at home and a little less confused in coming to such a large school.

3. Therefore, the author believes sin leads to social disorganization because Hebraism believes sin leads to complete disruption of obedience to the moral law which Hebraism believes leads to sound order.

4. Try to remember where you were last and whether or not you had the book at that time, and, if you cannot remember, try to think of someone who spent the day with you and a visit with that person may help you to recall events that will refresh your memory and give you a clue as to where the book is.

21   ¶ /no ¶

Paragraph Divisions

Paragraphing is a typographical convention used to aid the reader by marking the divisions of thought within an essay.

Usually each division of thought introduces a new topic. The topic of a paragraph is the word, phrase, or sentence by which that paragraph could be labeled or summarized.

A paragraph is often built around a topic. If a new topic is brought into the discussion, that fact may be indicated by a paragraph division, that is, an indentation. Here is an example of the end of one paragraph and the beginning of a new one from a theme on weather forecasting:

> . . . From the various cloud formations, then, a clever fore-caster can arrive at a pretty fair guess about tomorrow's weather.
>
> An even better basis for forecasting, however, is the pattern of atmospheric pressures. . . .

The most frequent errors are failing to indent and indenting too often. Failure to indent often results in a paragraph that seems to break right in the middle:

> In the history of man there have been different ideas about equality. There was absolute civil equality for the caveman. He had the same chance as his neighbor to kill a lion, or he could let the neighbor kill it and then try to kill the neighbor. The Greeks and Romans gave civil equality to those of noble birth. A patrician could be insane and have the vote, while an intelligent plebeian could not even voice an opinion. With the American Revolution, civil equality came to mean that all people were equal under the law. The idea of civil equality is still an ideal; we have far to go to make it a reality. Not everyone in our country can vote; not everyone has an equal chance for justice in the courts; not everyone has the opportunity for an education that will allow him to compete with others of equal talent; and we do not even pretend that everyone has the same social opportunities.

This paragraph develops two distinct though related topics. Thus the paragraph should be split in two:

> In the history of man there have been different ideas about equality. There was absolute civil equality for the caveman. He had the same chance as his neighbor to kill a lion, or he

could let the neighbor kill it and then try to kill the neighbor. The Greeks and Romans gave civil equality to those of noble birth. A patrician could be insane and have the vote, while an intelligent plebeian could not even voice an opinion. With the American Revolution, civil equality came to mean that all people were equal under the law.

The idea of civil equality is still an ideal; we have far to go to make it a reality. Not everyone in our country can vote; not everyone has an equal chance for justice in the courts; not everyone has the opportunity for an education that will allow him to compete with others of equal talent; and we do not even pretend that everyone has the same social opportunities.

The opposite fault, indenting too often, is perhaps more common. Beginning writers tend to paragraph every two or three sentences. Here is an example:

In preparing to read a play, one should first make a diagram of the stage and the properties it contains. This will be helpful in following the dramatic action.

Next, if any characters are described before the dialogue begins, one should make a mental picture of them. This will help to people the mind with living characters instead of puppets.

The third step is to glance quickly over the cast of characters, noting the main ones.

Now one is ready to begin the play.

All of these sentences develop a single topic, how to prepare to read a play, and should have been put into a single paragraph.

A third common error is introducing irrelevant material into a paragraph. All material ought to be related to the topic of the paragraph.

In terms of a statistical method or design, Gibson does not attempt any proof of his theory by testing or analysis. His article offers only speculation. There are no concrete results listed, only further speculation. *People should not be careless. Carelessness is the seed from which imperfection springs. This approach can ruin scientific research and make results almost meaningless.* Gibson feels that if his theory is correct, memory, which is the recalling of past images into consciousness, is an incidental accompaniment of learning and not its basis, as has been traditionally believed. Gibson lacks clarity. One does not write papers for learned journals carelessly defining terms and concepts. It is strange indeed that the article was even published.

REVISION PRACTICE

Try to make the paragraph indentations in the following passages coincide with the introduction of new topics. As you work, eliminate any irrelevant passages.

1. There were some problems during the first week of school. Getting oneself lost seemed to be the major difficulty; I think almost everyone had that problem at one time or another.

   I noticed also that many people had trouble in getting the proper books for the proper class.

   My biggest complaint was the manner in which things were done. Everywhere we hurried to get into line; then we stood and waited.

2. All knowledge, if it is to be considered scientific, must proceed from the scientific method. In order to use the scientific method the problem to be solved must first be defined, and this problem must be studied until a hypothesis, an educated guess, can be formulated as a solution to the problem. The scientist then must experiment to prove his hypothesis either correct or incorrect. If there are enough facts present to prove his hypothesis correct, then what is called a theory can be formed. A theory is the answer to the problem and remains as such unless proven by the facts to be wrong. A scientist is anyone who follows and uses the scientific method. The scientific method does not solve moral problems or set standards. Natural science includes

such fields as biology and psychology. Social science, on the other hand, is the study of people and civilizations. It involves all subjects concerning history, politics, and government.

22 PAC

Punctuation
of Adjective Clauses

Many students have difficulty in punctuating adjective clauses beginning with *who* (*whom*), *which*, or *that*. The difficulty arises from the fact that sometimes we use commas to separate the clause from the rest of the sentence, and sometimes we use no punctuation at all. The problem is to know when to punctuate and when not to.

First, here are three rule-of-thumb tests that will tell you when NOT to punctuate:

1. If the adjective clause begins with *that*, do not punctuate.

   The tie **that** he chose was blue and gray.

2. If you can substitute *that* for *who* (*whom*) or *which*, do not punctuate.

   The novel **which** I like best is *Tom Jones*.

3. If the *whom* or *which* may be omitted, do not punctuate.

   The dormitory (**which**) he lives in is Baker Hall.
   The club was highly pleased with the president (**whom**) they elected.

A useful way to approach the problem of punctuating adjective clauses is to examine their meaning. For example:

College students **who show initiative and responsibility** should be advanced rapidly.

Here the *who* clause limits *college students* to certain ones only. A limiting clause like this does not require commas.

College students, **who need relaxation as much as anybody,** should engage in social activities at frequent intervals.

In this sentence the reference is to ALL college students; there is no limitation. The *who* clause merely adds an independent descriptive detail. A *who* clause like this one, descriptive only, not limiting, requires commas. Now let us try two more:

The team **which wins the tournament** will be given a party by the losers.

Here the *which* clause points out which particular team out of all teams will be given a party. It limits the party receiver to one certain team. Here again we have a limiting clause, and no commas are used.

The Happy Hoboes softball team, **which has won 12 out of 13 games,** is given the best chance to win the tournament.

In this sentence there is no limitation by the *which* clause. This clause simply adds further information; hence commas are used. It may be useful to remember that after any proper name the *who* or *which* clause is nonlimiting and commas are used.

The following sentences are correctly punctuated. If you understand why, you should be in the clear.

The city that was damaged most by the flood was Omaha.

St. Paul, which is a manufacturing city, is the capital of Minnesota.

Those students who have not taken their health check must report to the Health Center.

We entered the Auditorium Building, which was being repaired, and tried to find the registrar's office.

Mary Moran, who lives in the next corridor, goes home nearly every weekend.

The girl whom I take to the formal must be a good dancer.

REVISION PRACTICE

The sentences below all contain adjective clauses beginning with *who* (*whom*), *which*, or *that*. They are all unpunctuated. Apply the tests that have been discussed above and put in the commas that are needed.

1. Everyone was there except Harry who knew the purpose of the meeting was hopeless.
2. Ticket orders which are not accompanied by checks will be ignored.
3. He who wants the most from his college education must develop regular study habits.
4. They will have little good to say of any person who is selfish and inconsiderate.
5. Although it was an unusually warm day, Avis who was wearing a heavy woolen sweater appeared cool and poised.
6. The belief that the pain will subside in a few hours is absolutely unfounded.

ADDITIONAL REVISION PRACTICE

1. Henry Cutler who has not missed a meeting in three years should certainly be a member of the committee.
2. These are the trails which have been designated for beginners.
3. Her broker who was supposed to be an expert cost her dearly within a month's time.
4. They were hunting the Mandarin goose which is found in Asia.
5. People who had long considered the move were at this point holding back.

6. Each lodge contained a color television set which raised the daily rate two dollars.
7. The next shock that I received was when I discovered the great amount of material available.

23 Paral

Parallel Structure

When a number of items are presented in a series, each item should have the same grammatical form. That is to say, each item should be a noun, or an adjective, or a prepositional phrase, or an infinitive, and so on. A passage having all members of a series in the same grammatical form is said to be parallel, or to have parallel structure. Parallel structure is desirable because it helps to make writing consistent, neat, and easy to follow. In each of the following pairs of examples, the first sentence contains a violation of parallel structure and the second sentence maintains parallel structure. As you study them, notice how the second sentence has been made parallel and how this parallel structure increases the ease in reading.

POOR: To make his organization clear, the speaker can blueprint his plan in advance, number his points, or many other things that will keep the audience on the road.

BETTER: To make his organization clear, the speaker can *blueprint* his plan in advance, *number* his points, or *do* many other things that will keep the audience on the road. [Series of verbs]

POOR: These three qualities I have discussed
—a friendly attitude toward others,
neatly dressed, and orderly management in the classroom—are the requisites of a good teacher.

BETTER: These three qualities I have discussed
—a friendly *attitude* toward others,
neat *appearance,* and orderly *management* in the classroom—are the
requisites of a good teacher. [Series
of nouns]

POOR: The principal of our school was required to teach subjects as well as
acting as administrator.

BETTER: The principal of our school was required *to teach* subjects as well as *to
act* as administrator. [Series of in-
finitives]

REVISION PRACTICE

1. The job of the agency is to examine returns, investigate them when necessary, and referring the violations
to the proper authorities.
2. The final step is to replace all of the chrome metal that
had been taken off and removing all the masking tape.
3. The article was divided into two sections for two reasons. First, to give the reader some facts and figures on
advancements in jet aviation; and second, the last section of the article was written for the purpose of informing the reader when and where the first jet aircraft was invented.
4. Some students coming to college are dependent on
others to see that they get places on time, look neat, and
various other things to be done.
5. The rooms will be much larger, decorations more elab-
orate, a tile bath, and plenty of wardrobe closets.

6. Four characteristics of a good salesman are these: neat appearance; capable of meeting the public; well informed on such subjects as world affairs, sports, and politics; and a firm believer in the product he sells.

ADDITIONAL REVISION PRACTICE

1. It is difficult to make a decision between sitting down to study or to go to a movie.
2. The refreshments should be simple, something all can eat, and gaily decorated and appetizing.
3. The stem of a watch has two functions—that of keeping the springs taut to insure constant running and used to set the hands if the watch runs slow or fast.
4. The five steps are as follows: (1) stating the problem, (2) unbiased observation, (3) hypothesis, (4) induction, and (5) conclusion.
5. If I were to classify the material I read, it would fall into two types, academic and pleasure.

24 Per

Periodic Structure

Let us look at two kinds of sentences—a short, solid one and a loose one. Here is an example of the first:

If you hire Smith, your office staff will greatly improve.

In this sentence we notice that, until we reach the very end, the structure is incomplete and thus the attention is held. This is called a *periodic sentence*.

The opposite of periodic sentence is a *loose sentence*. The loose sentence gets off to a good start, then wobbles and collapses weakly. Here is an example:

> Your office staff will greatly improve if you hire Smith, at least in my opinion, according to what I have heard.

This sentence would be structurally complete after *improve*. But after *improve* three successive increments have been tacked on, and the sentence dribbles away to a weak conclusion, with the less important details at the end—the position of emphasis. This is not to say that all loose sentences are bad. For example, the sentence above (the one beginning with *but*) is a loose sentence; although it is not immortal, it does carry its meaning effectively. The whole point is this: Do not write a loose sentence when a periodic sentence would be more effective. Let us examine a few loose sentences and note how they are tightened by revision into periodic sentences.

LOOSE: The stage looked dark and ominous at the opening of the curtain.

PERIODIC: At the opening of the curtain the stage looked dark and ominous.

PERIODIC: The stage, at the opening of the curtain, looked dark and ominous.

LOOSE: The college should provide better classrooms, many students believe.

PERIODIC: The college, many students believe, should provide better classrooms.

PERIODIC: Many students believe that the college should provide better classrooms.

LOOSE: A bargain is sometimes a poor investment for the housewife because people who put on bargain sales do not always carry quality merchandise.

PERIODIC: Because people who put on bargain sales do not always carry quality merchandise, a bargain is sometimes a poor investment for the housewife.

These illustrative sentences have been made periodic by the simple device of putting subordinate details at the beginning or in the middle.

REVISION PRACTICE

The sentences for revision that follow are not necessarily poor sentences, but you will learn a valuable lesson in emphasis by making them periodic.

1. Students at the college level try to avoid English 103 if it is at all possible.
2. We were concerned more and more with the deficiencies of the program as time went on.
3. The world would be a much better place to live in if people would first get rid of some of their major faults which stand in their way.
4. The lecture on a sociolinguistic approach to social learning was the high point of the evening in some respects.
5. Their leader became frightened and uncertain after he thought of all the punishments he had promised to bring to bear on the guilty men.

ADDITIONAL REVISION PRACTICE

1. Kurtz fell to his lowest depths through greed and through losing all contact with the civilized world, which had succeeded in keeping his baser emotions in check.
2. The loss of his billfold was the greatest tragedy, though the blowout and its ensuing delay accounted for our late arrival.

3. The intellectual must be capable of rendering sound and rational advice based on accurate reasoning, unlike the merely intelligent person who is concerned with the objective problems of life.

4. The well trained teacher is broadly educated, has a strong liberal arts background, and has an avid interest in current affairs according to John T. Laurence.

5. Camouflage was the most needed protective device, it seemed to us.

25 Poss

Possessive

It is sometimes difficult to decide whether to use the *of* possessive or the *'s* possessive. With animate things, either form may be used, though the *'s* is more common:

> The dog's leg
> The leg of the dog

But with inanimate things, writers of English tend to prefer the *of* possessive:

> The leg of the table
> The ceiling of the room

Since the possessive form of a noun is not differentiated from the plural form in speech, students sometimes forget to mark this distinction in their writing. A simple rule to follow is this. Write the word (whether singular or plural) as it is spelled without the possessive ending. If the word *as it stands* does not end in an s sound, add *'s*. Here are some examples:

WORD AS IT STANDS	POSSESSIVE
hero	hero's
knife	knife's
man	man's
men	men's
lady	lady's
Harold	Harold's

But if the word *as it stands* ends in an *s* sound, add either the apostrophe alone or *'s*, depending on which sounds the more normal to your ear. These examples will illustrate:

WORD AS IT STANDS	POSSESSIVE
heroes	heroes'
knives	knives'
France (ends in an *s* sound)	France's
Jones	Jones'
	Jones's
ladies	ladies'
boys	boys'

This rule also applies to compound pronoun forms (*anyone, anyone's; somebody, somebody's*) and to the pronoun *one* (*one, one's*). There is no apostrophe in the possessive forms of personal pronouns; for example,

Is this book his or hers?
Its fur was long and glossy.

REVISION PRACTICE

Add or delete apostrophes as needed. Change the *'s* to an *of* possessive when it seems desirable.

1. Jones book had been out over a year.

2. The sale of mens and boys clothing accounted for most of their financial success.
3. Heroes awards are exciting events at the time presented.
4. To everyones disgust he had forgotten his lines.
5. I am going to analyze Huxleys and Jungs views on religion, and by this method decide which authors viewpoints I would like to add to my own.
6. The storys ending was a happy one.
7. Its difficult to know its exact value.

ADDITIONAL REVISION PRACTICE

1. The families traits were all recorded in the experiment.
2. Iris purpose in answering the letter was obvious enough.
3. The companys success was determined by its dynamic leadership.
4. The rest of the report was completed on Nicks own time.
5. The ships sinking was the years important event.
6. Some teachers ask their students to correct each others papers.

26 Quot

Mechanics of
Quotation

Quotation marks are used primarily to mark words actually
spoken or written:

> He said, "It is unwise to consider any promotion at this time."
> The promise in your letter was to "love, honor, and obey."

Notice that the quotation is marked whether it is a complete
sentence or just part of a sentence; quotation marks indicate
that the enclosed words are those of the source, and not your
own.

An indirect quotation differs in that the material quoted is
not necessarily in the exact words of the source: therefore, it is
not marked:

> He said that it is unwise to consider a promotion at this time.

> or

> He thought that it would be unwise to consider a promotion
> now.

A *quote within a quote* is marked by a single quotation mark:

> She asked, "What is this 'science of the mind' to which the
> author refers?"

You should also refresh your memory on a few special uses of quotation marks.

First you should always underline (italicize, in printed matter) the titles of books, names of newspapers, periodicals, plays, movies, TV shows, operas, musicals; second, place quotes around the title of a work presented as a part of a book, such as a story, a poem, an article, an essay, a chapter:

> The material for this essay was found in "The Age of Absolutism," a chapter in Burns' Western Civilization.

Third, underline or place quotes around words used as words. These are words which do not play their ordinary role or which have some special meaning. For example:

> The word "word" is usually regarded as a noun.
> Is is a verb.
> Jefferson used "equality" and "liberty" as key words in the Declaration of Independence.
> We had argued for three hours about the meaning of literary.

Note that this situation calls for *either* quotes or underlining.

If you can remember the following three rules you will usually be able to handle the question, "Shall I put the punctuation inside or outside the quotes?"

1. The comma and period are *always* placed *inside* final or closing quotation marks.

> "Sing it brightly," he said.
> He said, "Sing it brightly."
> The conclusion of the essay is marked by the word "therefore."

2. It is always correct to place the semicolon and colon *outside* quotation marks.

> "Let your conscience be your guide"; with this remark he left the room.
> "Think big": this is the motto of the advertising world.

3. The question mark, exclamation mark, and dash are placed *inside* or *outside* the end quotation marks, depending

upon whether they punctuate the quoted words only or the entire sentence.

"How old is your sister?" he asked.

"Stop that noise!" she shouted.

Is this the answer I get, "No one is at home to you"?

What a stirring command, "Trust thyself"!

"The quality of mercy"—and by this Portia means human mercy—"is not strained."

Let us review one last problem. You have already noticed that a short quote (less than three lines) is simply incorporated in the body of the text; it may be introduced by a comma, dash, or colon, or it may appear as an independent sentence:

> We often find it best, in a discussion of literature, to begin with Aristotle's definition: "Tragedy, then, is an imitation of an action. . . ."

or

> We often find it best, in a discussion of literature, to begin with Aristotle's definition. "Tragedy, then, is an imitation of an action. . . ."

Long quotations (over three lines) are separated from the main body of the text by spacing and usually by indentation; they do not need quotation marks:

> Hazlitt spoke of a natural euthanasia many years ago when he stated:
>
> We do not in the regular course of nature die all at once: we have mouldered away gradually long before: faculty after faculty, attachment after attachment, we are torn from our-selves piece-meal while living. . . . The revulsion is not so great, and a quiet euthanasia is a winding-up of the plot. . . .

Three periods used here (. . .) indicate that words have been omitted at that point. The fourth period marks the end of a sentence. This is a convenient device for cutting out irrelevant material in quotations, so long as you do not disturb the structure or meaning of the sentence.

REVISION PRACTICE

1. He said, I'll not bring up the matter again. This is final. I want to hear no more about it.
2. Is this article Cognitive Dissonance, Sensitivity and Evaluation, she asked, to be found in the Journal of Abnormal Psychology or in the Scientific American?
3. Who is the King of Glory asks David?
4. I didn't use the word kinesthetic but the word kinetic.
5. He said that, "our technology tends to put us back into the booming, buzzing, humming confusion of infancy and primitive tribal living."
6. These same people ask where would we be today if there were no writing.
7. They are always the same old questions who would do the work if I weren't here? Who would pay the bills? Who would listen to your troubles?

ADDITIONAL REVISION PRACTICE

1. Haste makes waste: this was her motto and it resulted in her continually being late for everything.
2. The words of the registrar boomed through the auditorium. Registration is the process of signing up for classes. New students receive advice from faculty members in planning their schedules, and all students fill out cards and material for the use of various offices on campus. The last step in registration is payment of bills, and the student is not registered until financial arrangements have been completed. Information about courses and fees is contained in a publication with the title, The Academic Bulletin. Students may pick up their registration materials Monday morning.

3. Have you read this week's Time, he asked?

4. I now asked myself the question what is "true science?"

5. While serving as an editorial writer for the Chicago Daily News, Sandburg wrote such poems as Good Morning America, and Broken-faced Gargoyles.

6. How long are you going to wait, he was asked.

27   Red

Redundancy

Redundancy refers to a needless duplication of meaning; for no good reason, the same idea is repeated in different words. Redundancy frequently takes such a form as "basic fundamentals," where the meaning of the adjective is included in the meaning of the noun. Thus, redundancy always leads to wordiness (*see* Section 40 W). Both result from carelessness and haste and should receive particular attention during the revision process. Let us reduce these sentences to their essentials:

POOR: She had the radio on so low it was not audible to the ear.

BETTER: She had the radio on so low it was not audible.

POOR: Eighteenth-century England was ruled by a monarchy type of government.

BETTER: Eighteenth-century England was ruled by a monarchy.

REVISION PRACTICE

1. The house is low and rectangular in shape with a sloping green roof.
2. He may notice that the sky is a bluish gray in color.
3. This program gives a person a period in which to adjust to the new surroundings that encircle him.
4. Things will be easier in future days to come.
5. The area used to be covered by water in the past.

ADDITIONAL REVISION PRACTICE

1. In conclusion let me close with this word of advice—cool it!
2. We have created antibiotics to rid ourselves of germs already in us.
3. You'll notice that the rose is a dusty pink in color.
4. Machines often allow people to idle away time which is usually very carelessly wasted.
5. In my opinion, I think there is good in both.

28 Ref

Reference of Pronouns

The pronouns most likely to cause trouble are: *it, they, this, who,* and *which.* In using such pronouns, the writer should keep two points in mind. First, he should make sure that singular pronouns refer back to singular words and that plural pronouns refer back to plural words. This seems simple, yet freshmen often write sentences like these:

> Whenever a *student* needed advice, *they* could feel free to go to Mr. Maire.
>
> As long as the public wants better *motion pictures,* the motion picture industry will supply *it.*

Second, the writer should make sure that the reader knows exactly to what the pronouns refer. Let us examine a few illustrative sentences from freshman writing:

> Men like Brodie and Kolmer discovered vaccines and gave them to the public, but *they* were not successful. . . .

Here the reader cannot tell whether *they* refers to *vaccines* or to *Brodie* and *Kolmer.* The difficulty can be easily remedied by deleting *they* and inserting words that say exactly what the writer means:

> Men like Brodie and Kolmer discovered vaccines and gave them to the public, but these vaccines were not successful. . . .

or

> Men like Brodie and Kolmer discovered vaccines and gave them to the public, but these men were not successful. . . .

The pronouns *which* and *this* can be a special source of confusion because they may refer to a whole preceding idea or to a particular noun. In the next sentence the reference of *which* is not immediately clear:

> Each room has two study desks with only one study lamp, *which* is very inconvenient because of the poor lighting on one of the desks.

To revise this unclear reference of *which,* the writer has only to substitute a word or more which says exactly what he means:

Each room has two study desks with only one study lamp. This situation is very inconvenient because of the poor lighting on one of the desks.

REVISION PRACTICE

1. Sometimes a student causes trouble simply because they have not learned what is right and what is wrong.

2. Television programs are interrupted every now and then for commercials. Some of them are very irritating.

3. School spirit is low because students are not acquainted with one another. This makes it hard to get backing for the teams.

4. When Columbus heard that Bobadilla had in his possession a letter from the sovereigns giving him the power he was exercising, he meekly went to confer with Bobadilla.

5. He was always bringing into our room some strange dog he had found, which was a nuisance when I was trying to study.

6. These services are very worthwhile for it enables the student to actually take part in the worship.

7. Fraternity houses on the university campus would not only provide more room but also guidance, social functions, and the development of leadership for its members.

8. The dormitory rooms have built-in closets which have eight drawers and two shelves. They have a place to hang clothes and a sliding-door cabinet to place boxes and suitcases in.

9. The rooms are heated by a radiator which is run by steam. They can be easily regulated by the student with the temperature desired.

ADDITIONAL REVISION PRACTICE

1. Kindergarten is the child's first experience at school and they are very excited about learning.
2. Each member of the faculty had at least one school activity which they were to sponsor.
3. There are many things that are needed to be a good wrestler. They must have a knowledge of the various holds and escapes.
4. Band means a lot to a person and certainly does help them.
5. This teacher was mathematically inclined but was unable to get it across to us students.
6. The Roman farmer's occupation was closely related to his religion. They would pray to the various gods for help.

29 Rep

Repetition

The symbol *Rep* indicates that a word has been repeated unnecessarily and thus draws unwarranted attention to itself. This is a matter of style rather than of clarity. In the following sentence, for example, the second instance of *spring* could easily be avoided by substituting *then* for *in spring:*

> *Spring* is the time of year I like most, for *in spring* we hunt in the woods for *spring* flowers.

Repetition in itself is not a writing fault. It is very useful in achieving coherence, as you will learn in the section "Achieving Sentence Flow." It is also a device for emphasis, as the following sentences show:

> I *hated* the work, I *hated* the hours, I *hated* the job.
> Our subject will be *patriotism:* not the cheap *patriotism* of the stump politician, but the *patriotism* of the conscientious citizen.

A simple way to discover unnecessary repetition is to read your written material aloud. To eliminate the repetition, replace the repeated word with a synonym or pronoun, or omit it altogether.

REVISION PRACTICE

1. He thought that this class thought it had no responsibility to uphold tradition.
2. I cannot explain what it is that makes me feel like a new man. It could be this feeling of self-confidence in my ability to be able to work hard for a good average that makes me feel that my prospects are good.
3. He did not list the important steps in order of their priority in his list of ideas.
4. The pond on the campus adds much beauty to the campus.
5. He had to report for duty at least three weeks before anyone else had to report.

ADDITIONAL REVISION PRACTICE

1. Because of their reaction, I felt all along that they felt this way about including Jim in the party.
2. He believed that many people believed the same story he did.

3. The furnishings with which the house was furnished were in excellent taste.
4. All these points would seem to directly point to the matter at hand.
5. I feel that new students' week was invaluable to those students who took advantage of all the advantages given them.

30 RO

Run-on (Fused) Sentence

A run-on, or fused, sentence is one that follows a preceding sentence as if both were one. There is no period and no capital letter to show the break between them (*see* Section 6 CF). Some run-on sentences do not interfere with meaning, but they suggest that the writer is illiterate:

> Our lounge is the Rose Lounge it is very beautiful.

Others, however, leave the reader in doubt about the intended meaning:

> Of course we had to go shopping the very first day we discovered more of the virtues of this lovely city.

The preceding example might be read as one sentence, but what the writer meant was:

> Of course we had to go shopping. The very first day we discovered more of the virtues of this lovely city.

For further help with this problem *see* Sections 32 S Col and 36 Sub.

REVISION PRACTICE

1. No matter what type of subject matter you choose, your report on a book with ideas should of course emphasize the ideas you are required to know what these ideas are.
2. As you are completely new in this business you are allowed certain mistakes too many, however, might cost you your job.
3. A person's character is acquired from his environment he is not born with it.
4. No single text was assigned the students found explanations in different books without exception these books supplied different answers.
5. Scholarship is recognized at this university through the Dean's honor designations, scholarships, awards, and prizes a person with good grades is rewarded.
6. Was this the only solution to the dilemma it seems so.
7. This assumption fails to consider the fact that not all information is equally useful it is therefore not entirely useful.
8. Rowles is married to the former Virginia Davis they have three children.

ADDITIONAL REVISION PRACTICE

1. She is entitled to think whatever she likes about my paper her arguments would have been more convincing if she had read it however.
2. I fervently hope you will make your voices heard beyond that I dare expect no more.
3. Think what you like I don't care.

4. Desmond misunderstands the character of the mistaken U.S. intelligence estimates with which the book is mainly concerned he engages in a useless and pointless battle with a strawman.

5. While hopes for diplomacy are being dashed, the arms race has mounted the long range outlook is less than sanguine.

6. It was an idea not too difficult to put into practice on that basis it was quickly accepted.

7. Was all this hard work and studying of rudiments worth it I think so.

31 SB

Sentence Beginning

The symbol SB means that you should begin your sentence in a different way. Your instructor may indicate the way he recommends by using a number after SB. SB9, for example, would mean the ninth sentence beginning—the infinitive—among the twelve kinds listed below.

There are at least a dozen grammatical ways to begin a sentence. If you have an active command of these twelve, it will be easier for you to start a sentence and to find ways of connecting it in thought with the preceding sentence. The most frequently used ways are these four:

1. The subject—

Jim was afraid of the impending examination.

2. The adverb—

Unfortunately, Jim has not kept up with his assignments.

3. The prepositional phrase—

Before the impending examination, Jim realized that he had neglected his work.

4. The subordinate adverbial clause—

Because he had been negligent of his assignments, Jim was fearful of the impending examination.

The remaining kinds of sentence beginnings are more characteristic of writing than of speech, and you will find it worthwhile to master them. This is easy to do and it will pay handsome dividends in increasing your writing skill.

5. The direct object—

His daily assignments Jim had neglected for several weeks.

6. The adjective—

Fearful of the impending examination, Jim began to study in earnest.

7. The present participle—

Worrying about the impending examination, Jim stayed up all night to study.

8. The past participle—

Deprived of study time by play practice, Jim found himself unprepared for the impending examination.

9. The infinitive—

To keep up his grade average, Jim kept faithfully to his study schedule.

10. The absolute construction—

Examination time approaching, Jim set aside an hour a day for review.

11. The coordinating conjunction (*and, but, for, or*)—

Jim wanted to participate in the Homecoming festivities and try out for the leading role in the first play of the drama workshop. *But* examination time was near.

12. The appositive before the subject—

A playboy during the semester, Jim now had to face the ordeal of final examinations.

REVISION PRACTICE

Revise the following sentences using the sentence beginnings suggested in the parentheses.

1. Frank made several hurried changes in plans in order to meet Ginger at the hour they had set. (Infinitive)
2. Bill Pace was a contender for class honors. Yesterday he failed the midterm examination. (Appositive)
3. He was relieved of his duties immediately and left his position under a cloud of suspicion. (Past participle)
4. Hoping for an agreement on a rate increase, the board of directors called an early meeting. (Subject)
5. The old plane, which was slow and noisy, was used for short runs and for emergencies. The new one was used for daily passenger service. (Coordinating conjunction)
6. He uttered an enthusiastic exclamation and left the room. (Present participle)
7. His difficulties were resolved during the coming months. (Prepositional phrase)
8. The committee had not planned the program of the annual art festival early enough to get the services of well-known speakers. (Direct object)
9. John had probably insisted on an early payment. (Adverb)
10. Harry was indignant about the noise in the library and wrote a letter of complaint to the *University Daily*. (Adjective)
11. Barry had the driver in his hand and stood watching the young couple trying to get off the first tee. (Absolute construction)
12. New members had to be appointed, for the safety committee had been negligent in its duties. (Subordinate adverbial clause)

32   S Col

Use of Semicolon
Recommended

A writer should be familiar with the three basic uses of the semicolon.

First, the semicolon is used between independent or main clauses to indicate a separation stronger than that given by a comma but less strong than a period stop.

> Harris was always generous with his children; in this respect he was like his father.
>
> The comma indicates a slight pause; the semicolon is used for a stronger one.

Second, the semicolon is used between independent or main clauses when the second clause begins with a transition word such as *therefore, hence, then, moreover, also, accordingly, however, thus, otherwise.* It is modern practice to use no comma after the transition word, though one may be used.

> He had asked several times; hence he felt it unnecessary to repeat his request.
>
> We finished adding the figures; then we counted the money.

The transition word may come later in the second clause.

> The light on the boathouse flooded the landing area; he felt therefore that the boats were safe from theft.

Third, the semicolon is used to separate clauses or phrases in series when commas are used within these clauses or phrases.

> Replenishing the stock of the wholesale house required particular attention in the purchasing of these items on specific days: apples, oranges, and bananas on Monday; beets, radishes, and celery on Tuesday; milk, wine, and soda pop on Friday.

Experienced writers frequently use the semicolon as a strengthened comma to emphasize a pause.

> The community needs a strong man; that is, one with special qualifications.
> The nose may be shut off from the mouth by lifting the soft palate; or, if the soft palate is allowed to hang down freely, these make a combined resonance chamber.

REVISION PRACTICE

1. It is impossible to clarify the issue at this time, however, in a month or two much of the research will be in and facts not known at this time will be readily available.
2. When we are under stress we do not write as well, thus we have a harder time with an in-class theme.
3. The catalog had made it quite clear in order for a person to graduate he must have at least 128 semester hours.
4. The long grass, the broken windows and unpainted exterior all suggested a house with no occupants, therefore it seemed surprising to observe smoke coming from the chimney.
5. The plane was to be several hours late, he knew therefore that the flat tire was not going to create an emergency.

6. Hardly anyone knew how the rumors had started, the whole town was nonetheless aware of them.
7. The following people have consented to furnish letters of reference: Mr. James C. Connor, Manager of Pen Department, Chandlers Inc., Evanston, Mr. Henry Trauscht, owner, Evanston Auto Parts and Supply Co., Evanston, Mr. Szmzayk, Superintendent of Mails, Evanston Post Office, Evanston.

ADDITIONAL REVISION PRACTICE

1. He had turned in his records over a month ago hence he felt no obligation to write a letter of explanation for any delay.
2. Other books by Hardy are listed for you with the dates of publication: *Far from the Madding Crowd*, 1874, *The Return of the Native*, 1878, *The Mayor of Casterbridge*, 1886.
3. His real interest lay in his business, he devoted as much time as he could however to the study of history.
4. After cutting the wood for the fireplace he was exhausted and decided to take a nap, consequently he didn't hear the telephone.
5. I checked the directory faithfully to find the Anderson Company, which was to do the plumbing, Clifton, who was to do the wiring, Bottleson, who was the recommended cabinet maker, and Graber who was the interior decorator.

33 Sl

Slang

Although slang terms are often vivid and colorful and sometimes seem to express exactly what you mean, they have no place in the formal style used in most college writing. There is always an equivalent for a slang term in the standard vocabulary. It is not acceptable to use slang terms even with apologetic quotation marks, as in "He 'goofed off' in the last examination." For *goof off* there is a wide array of substitutes, most of which you know: *blunder, botch, fumble, err, mismanage, make a gaffe* or a *faux pas,* and others.

Slang cannot be defined with precision, but we can make a few descriptive statements that will help to identify slang words:

1. Slang is lively and picturesque.
2. It is usually short-lived, being here today and gone tomorrow.
3. It varies with different speech communities. High school slang differs from college slang, and British slang from that in the United States.
4. It is found more in speech than in writing.
5. It is made up of certain kinds of words:
    a. Clipped words—*prof, lab, flunk*
    b. Newly invented words—*pizzaz*
    c. Words borrowed from jargon, that is, from the specialized vocabulary of a profession, trade, or similar restricted group—"Let's go to your *pad.*"

    d. Old words with new meaning—"He is a *jerk*," "Don't be *chicken*."

    e. Standard words used figuratively—"They had a *blast* last night."

34 SP

Shift of Person

The word *person* is here a grammatical term. Pronouns that refer to the speaker, like *I* and *we*, are said to be in the first person. The pronouns *you* and *your*, which refer to the person spoken to, are called the second person. Other pronouns and all nouns are called the third person. For example, *they, it, he, she, one, a person, college, writings* are all in the third person.

    The writer must be consistent in his use of grammatical person and not shift aimlessly from one person to a different person. The most common mistake with grammatical person is to shift without cause to *you* in the middle of a passage that is written from the first-person point of view (*I* or *we*) or the third-person point of view. Here is an example of this common mistake:

> It is hard for me to write because I can never think of the right words that will best express what you are trying to put down on paper.

It is obvious that the *you are* should have been *I am*. This was a shift from the first to the second person. In the next example the shift is from the third to the second person:

Meeting new people broadens a person's train of thought and develops your personality.

This might have been written in either of two ways. If the writer had been addressing the reader directly in the paper, using the second person *you*, then the sentence might have been written:

Meeting new people broadens your train of thought and develops your personality.

If, on the contrary, the writer had been using the third person, then he should have kept this person throughout:

Meeting new people broadens a person's train of thought and develops his personality.

REVISION PRACTICE

1. In high school I liked grammar because it was easy for me, and you could grasp it on your own.
2. When we had received our bedding, we were shown how to make one's bed according to army regulations.
3. When a person enrolls for a two-year course, he must follow a prescribed schedule set up by the college. The reason for this is that you will get the classes that will be most helpful to you as a teacher.
4. I need to understand what good writing is. To be able to express yourself in writing is of great importance.
5. College, to me, marks the beginning of a new life, a demanding life, from which you can expect to receive only as much as you give.
6. I liked this course better than my first year's work. You were more on your own in using the library and finding references.

ADDITIONAL REVISION PRACTICE

1. Many of these writings I thoroughly enjoyed, and they offered an extra advantage in that they taught one to read properly.
2. I wanted to find out all the rules I needed to know to keep you out of trouble.
3. By the end of one's first year, you feel you know just about all the answers.
4. He believes that all man needs to do is to get all the facts and you will get the correct answer.
5. I like being out on the prairie where you can look for a mile in any direction.

35 Sp

Spelling

The symbol *Sp* represents one of two writing faults:

1. An actual misspelling, like *proceedure* for *procedure*.
2. A choice of the wrong word, as "We may *loose* the game," instead of *lose*.

If your fault is an actual misspelling, you may find the word on the list immediately below, which consists of words that college freshmen frequently misspell. If it is not there, you will have to look it up in your desk dictionary.

If the *Sp* represents a wrong-word choice, you are likely to find your incorrectly used word on the second list below, together with the word you should have used.

WORDS OFTEN MISSPELLED

accommodate	dependent	loneliness	quantity
achievement	description	losing	realize
acquire	disastrous	marriage	really
all right	embarrass	Negroes	receiving
analysis	environment	noticeable	referring
analyze	equipped	occasion	repetition
apparent	especially	occur	rhythm
argument	essence	occurred	sacrifice
beginner	excellent	omitted	sense
believe	existence	original	separate
benefited	experience	performance	shining
business	explanation	philosophy	similar
category	familiar	possession	studying
changeable	fascinate	precede	succeed
choice	forty	predominant	surprise
chosen	fourth	preferred	tendency
comparative	government	prejudice	theories
conscience	grammar	prevalent	thorough
conscious	height	privilege	tried
consistent	imagination	probably	unnecessary
controlled	immediately	procedure	usually
controversy	intelligent	proceed	useful
criticism	interest	prominent	varies
decision	interpretation	professor	various
definite	led	pursue	writing

WORDS OFTEN CONFUSED

advice	Give me some advice.
advise	I'll be glad to advise you.
affect	Will it affect my grades?
effect	He studied the effect of fatigue.
accept	Will you accept the money?
except	Everyone except Sam was there.
choose	Be sure to choose the best.
chose	Yesterday she chose a dress.
cloths	She covered the bread with cloths.

clothes	Clothes make the man.
complement	The tie should complement the shirt.
compliment	Do not forget to compliment your hostess.
it's	It's a nice day.
its	Its head is small.
lose	They lose their lands.
loose	The boat broke loose from its mooring.
passed	He passed the test.
past	The past few days were busy.
principle	Honesty is a good principle.
principal	The principal was in his office. Their principal food was rice. The principal was $12,000.
personal	I have personal business.
personnel	Many personnel were injured.
quiet	It was quiet in the house.
quite	It was quite dark outside.
role	Sam will play the role.
roll	He called the roll.
site	We shall visit the building site.
cite	Please cite an example.
than	He is heavier than I.
then	Then he left his room.
their	Their house is white.
they're	They're both my friends.
there	There is a boat at the dock.
two	He has two cars.
too	I want some too. It is too late.
to	They went to college.
weather	The weather is sunny.
whether	He did not know whether to go or stay.
who's	Who's going to the game?
whose	Whose book is this?

36   Sub

Subordination

The symbol *Sub* indicates that you should change an independent clause into a subordinate element.

POOR: George felt exhausted after the long exam, and he decided to relax at a light movie.

BETTER: Since he felt exhausted after the long exam, George decided to relax at a light movie.

George, who felt exhausted after the long exam, decided to relax at a light movie.

Feeling exhausted after the long exam, George decided to relax at a light movie.

Exhausted after the long exam, George decided to relax at a light movie.

Subordination is a matter of writing style about which there are no absolute rights and wrongs. The general principle is this: Put the important part of your message in an independent clause and the supporting or collateral parts in subordinate elements.

A subordinate element of the kind needed for this sort of correction usually falls into one of the following four categories:

1. A clause beginning with a word like *as, since, while, because, when, although, if, unless, after, before.*
2. A clause beginning with *who, whom, whose, which,* or *that.*

3. A phrase beginning with an *-ing* verb.

4. A phrase beginning with an *-ed* verb or an adjective.

REVISION PRACTICE

The sentences in the following revision exercises are not necessarily poor ones, but by revising them you will gain skill in manipulating subordinate elements.

1. Most people are afraid of their first driver's examination, and they are unable to take it with any degree of confidence. (Begin with the adjective *afraid*.)

2. Monsieur Vaillot was a lawyer and he knew exactly what to say to the other driver. (Do this two ways, using *because* and *who*.)

3. One must report regularly and on time or his entire career may be in jeopardy. (Use *unless*.)

4. We filled our glasses and then drank a toast to absent friends. (Use *after*.)

5. The half-time show was much too long, so the opening of the second half had to be delayed ten minutes. (Do this in two ways, using *since* and an *-ing* verb.)

ADDITIONAL REVISION PRACTICE

1. You must have your central thought in mind and then there are a number of excellent ways to develop your paragraphs.

2. This article was very clear and readable and the author used terminology that even the layman could understand.

3. His paper has been published in the college literary magazine so one can assume it has some merit.

4. I finally located the proper street and then headed for the nearest pub.

5. My friend had left his billfold at home and so I provided the funds for the evening.

37 SV

Sentence Variety

A series of sentences following the same pattern tends to become monotonous. The passages below exemplify this problem. Each is followed by an improved passage showing an acceptable solution.

Subject-verb pattern   The team quickly lined up for the next play. The quarterback crouched low. He barked out the signals. The ball was snapped to him. He wheeled to the right and feinted a pass. Then he plunged straight through center. He slipped at the line of play and could not regain his feet. So he was downed without a gain.

Improved   Quickly the team lined up for the next play. Crouching low, the quarterback barked out the signals. When the ball was snapped to him, he wheeled suddenly to the right, feinted a pass, and plunged straight through center, but slipped just at the line of play. Unable to regain his feet, he was downed without a gain.

Present participle beginning   Having heard a scratching noise at the lake shore, Pete stumbled out of the tent and peered through the darkness. Having bought a brand-new canoe, he was worried about someone's stealing it. Seizing his flashlight, he ran down the path toward the boathouse. Turning the light on the canoe, he saw a fat coon scuttle away.

Improved   Having heard a scratching noise at the lake shore, Pete stumbled out of the tent and peered through the darkness. Because his canoe was brand-new he was worried about someone's stealing it. He seized his flashlight and ran down the path toward the boathouse. As he turned the light on the canoe, he saw a fat coon scuttle away.

REVISION PRACTICE

Rewrite each passage to eliminate the monotony of grammatical pattern.

1. **Present participle beginning**   Having finished his class work for the following day and having no exciting plans for the rest of the evening, Frank decided to stay home. Browsing through his limited library, he found nothing at all he had not already read. Yawning and groaning at the same time, he decided to go to bed.

2. **Passive voice pattern**   The course is intended to introduce freshmen in the humanities course to Italian art. The history of art forms is studied from the Early Renaissance through the Baroque Period. Styles of painting and sculpture are analyzed. The gradual development of realism is examined. The principal works of major artists are used as illustrations by the instructor.

3. **Adverbial clause beginning**   Since Norma was late this morning, she found herself running toward the bus stop. When she turned the corner, she saw her bus pulling away from the curb. After she had waved frantically at the departing bus and had lost her breath

chasing it, she sat down on the curb and burst into tears.

38 T

Transition

Sentences should flow smoothly, one into the other. One way to achieve this flow of thought is to make sure that the first part of a sentence refers back to the preceding sentence. (Other ways are discussed under "Achieving Sentence Flow" in Chapter 3.) When there is a gap in thought between two sentences, your instructor may use the symbol *T*. This indicates that you need some word or word-group to bridge the gap.

Sometimes all you need to do is add a transitional word or phrase such as *next, besides, then, however, thus, consequently, furthermore, indeed, in addition, therefore, moreover, hence, nevertheless, on the contrary, as a result.* This passage will illustrate:

> My adviser was the first person I consulted. He suggested that I drop one course. I went to see the dean, who willingly consented to the change.

Here the addition of a single word will effect a smooth transition:

> My adviser was the first person I consulted. He suggested that I drop one course. Then I went to see the dean, who willingly consented to the change.

At other times you may need to revise the second sentence so that the first part of it refers back to the previous sentence. Here is an example:

> Harris tore a ligament in his shoulder at the first tennis match. He was unable to play for a month after that.

In this case a change in the position of one phrase will make the transition:

> Harris tore a ligament in his shoulder at the first tennis match. After that he was unable to play for a month.

In the next example the second sentence must be readjusted and a clause inserted to make the transition:

> Johnson worked day and night to get his report ready before the deadline. The president praised him for his conscientious effort when the report was finished and turned in.

Here it is again, with the first sentence flowing into the second:

> Johnson worked day and night to get his report ready before the deadline. When the report was finished and turned in, the president praised him for his conscientious work.

REVISION PRACTICE

1. After five miles of steady marching with full field equipment, the company halted in the shade of a small grove. They rested for fifteen minutes there.
2. My roommate's car broke down just as he was returning to the campus on Sunday evening. He was late to his morning class because of this.
3. During the first two weeks each student delivers four five-minute talks. An introduction of himself constitutes the first of these.

4. In the nineteenth century writers tended to write long, complicated sentences. They ordinarily used much more punctuation than is common today, in order to guide the reader through the intricate patterns of these long sentences.

5. My roommate spent most of his evenings at a nearby tavern. His semester grades were low as a result.

ADDITIONAL REVISION PRACTICE

1. I would say on first reading that these are the views of a pessimist. I find that this is not so after several readings.

2. I returned to the dorm late. I was campused a month for the misdemeanor.

3. Jim dubbed his first shot off the tee. He could not do anything right after that.

4. The play proved emotionally disturbing. They discussed it until midnight after they left the theater.

5. The class president announced the names of yearbook committees. He congratulated the committee members and outlined their duties.

39 Te S

Tense Shift

A writer must be careful in his use of the tense forms of verbs to show time. When he does not select the tense form which would indicate the precise time relationship that he has in mind, or when he shifts from one tense to another without reason, he may confuse his reader or muddle the time perspective in his writing.

One problem occurs in writing about events of the past:

> It *was* Jefferson's belief that the safety and the freedom of the colonies *is endangered* by England.
>
> Sandra *is offered* her first job in the second chapter. Three chapters later she *accepted* it.

You may present such events either in past or present tense (the latter is called *historical present*), but you must choose one tense or the other and stick with it.

> It *was* Jefferson's belief that the safety and freedom of the colonies *were endangered* by England.
>
> Sandra *is offered* her first job in the second chapter. Three chapters later she *accepts* it.

It is also easy to misuse tenses in indirect quotations. For example, in using Samuel Pepys' statement, "This play is the most licentious to come on the stage all year," a student may carelessly write:

> Samuel Pepys *said* that this play *is* the most licentious to come on the stage all year.

A safe rule to follow in such situations is this: Use past tense in the indirect quotation when the main verb of the sentence is in the past tense.

> Samuel Pepys *said* that this play *was* the most licentious to come on the stage all year.

But use the present tense when the meaning of the quotation is true in the present.

> Francis Bacon sincerely *believed* that reading *makes* a full man.

REVISION PRACTICE

1. While you were looking the other way, a miracle happens. Without even noticing, you made the green light at Third Street.
2. Just when he thought the work was finished, he finds another problem to do.
3. She seemed utterly uninterested in the contest when suddenly she jumps up and screams, "Sock 'em!"
4. The church believes that we will never lose our moral standards because so long as we believed in a God we can regard Him as the source of our moral ideas.
5. At the end of her story she got her facts all mixed up; she is sure, however, that no one doubts the truth of her tale.

ADDITIONAL REVISION PRACTICE

1. Giovanna allowed Frederigo to spend all his money on her; leaving him after it is gone is selfish on her part.
2. The author calls for a religion that makes life simple, clean, beautiful, and gives a high standard to men. He wanted a religion like this because he wanted to think of himself as being highly intelligent.
3. Then I ask if she wants to go out with me but she brushed me off with a look and left me standing there.
4. Finally, as I got all my clothes situated and bought my books, I think how confusing this week at college would have been.
5. Thus we can see that religion has evolved, or at least seemed to have evolved.

40 w

Wordiness

Writing should be direct and spare, free from unnecessary words which serve no purpose. When you convey your thoughts with an overabundance of words, your style seems bloated and your reader is likely to become impatient. A first draft is often wordy, but when you revise you can pare away the unneeded expressions and try to achieve an economy of style consistent with accuracy of statement. Notice how a little delicate surgery improves the following sentences.

WORDY: In many schools they have regular film operators.

BETTER: Many schools have regular film operators.

WORDY: I could also make some improvements in my articulation of words.

BETTER: I could also improve my articulation.

WORDY: I slur over words for the simple reason that I have become lazy.

BETTER: I slur over words because I have become lazy.

WORDY: I think my speech would have been better if I would have gotten rid of three weaknesses. The three weaknesses I have are the tempo in which I speak, the articulation of my speech, and the spurtiness of my speech.

BETTER: I think my speech would have been
better had I got rid of three weaknesses: my fast tempo, slovenly articulation, and spurtiness.

For additional exercises *see* Sections 20 OBS and 27 Red.

REVISION PRACTICE

1. If you want to find the latest styles you have to go to
another town that is larger.
2. In order to achieve good grades in school some people
have to cheat for them.
3. As far as extracurricular activities go, there is a wide
variety of intramural and intercollegiate sports available.
4. The other rate which is bad for the speaker is the rate
which makes the speech too slow.
5. College offers many new opportunities for students,
such as the following: there are countless activities
that students can engage in, and there is knowledge
that students can gain from their classes.
6. Dennison has many changes that should be made in
order to be a better town.
7. Without these qualities those who study to be teachers
will be nothing but failures, as far as teaching goes.

ADDITIONAL REVISION PRACTICE

1. People who have had little or no experience in the field
of athletics often think that it takes only brawn and
few brains to make a basketball player.
2. She did not realize the fact that a student studies best
when he is fresh.
3. One thing that I can say is that I never attended as
many dances as I did in that week.

W   40

4. In conclusion, generally then, Thoreau's essay, I believe, was written to tell the people how to remove themselves from the power of the government for their own betterment.

5. I intend to discuss the male students in this college. This group of almost five hundred men is composed of freshmen, sophomores, juniors, and seniors. They tend to fall into two main classifications: the studious students and the nonstudious students.

6. Each student has his own preference as to how he should divide his time between scholastic and social activities.

7. I would like to point out a case where the views of the church and the views of science are similar. The case in point is original sin.

41 ww

Wrong Word

We sometimes confuse words because of their similarity in spelling, or pronunciation, or meaning. The following checklist gives pairs of words that students are likely to use in the wrong way. (Words which are pronounced the same, but which differ in spelling and meaning are called *homonyms*.)

accept / except	"*Except* for the fat ones, we will *accept* the lot."
affect / effect	*Affect* is usually a verb (to influence), *effect* usually a noun (the re-

sult). "The *effect* of the war on him was slight." "Wool *affects* his sinuses." *Effect* also appears as a verb. "The new Party promises to *effect* (bring about) reforms."

allusion / illusion
Illusion (a false impression) is often used for *allusion* (a reference). "He made an *allusion* to Lady Macbeth's *illusion* of blood on her hands."

amount / number
Amount refers to uncountable quantities, *number* to countable quantities. "They bought a large *amount* of silk and a *number* of paintings."

born / borne
(Homonyms) *Borne* is a form of *bear* (to carry). "The child is *born* in the hospital and *borne* home by his parents."

censure / censor
Censure (to condemn, blame), *censor* (to expurgate, or remove offensive sections). "The chief was *censured* for *censoring* our mail."

cite / site
(Homonyms) *Site* (location) is often substituted for *cite* (to refer to). "He *cited* the case of the new tailor shop; its *site* was unsuitable."

contemptible / contemptuous
"A cruel man is *contemptible;* a *contemptuous* man is haughty."

continuous / continual
Continuous (extended without interruption), *continual* (repeated often). "*Continuous* service, thought, siege, woods." "*Continual* dripping."

council / counsel
(Homonyms) A *council* is a group of advisers; an adviser *counsels,* or gives *counsel.*

credible / credulous
"*Credulous* people believe anything they hear, regardless of how *credible* (or incredible) it might be."

disinterested / uninterested	Careful usage distinguishes between *disinterested* (not prejudiced) and *uninterested* (not interested). "The judge is a *disinterested* listener." "I am *uninterested* in your plan."
fewer / less	*Fewer* refers to number, *less* to amount. "He is smoking *fewer* cigars, and drinking *less* beer."
human / humane	All men are said to be *human;* kind men are said to be *humane*.
imply / infer	*Infer* (to draw a conclusion from evidence) is often used for *imply* (to hint). "I *implied* a compliment, but from my words he *inferred* an insult."
ingenuous / ingenious	*Ingenuous* (frank, without guile), *ingenious* (clever, inventive). "An *ingenious* man can take advantage of an *ingenuous* one."
its / it's	*It's* is used only for the contraction of "it is." *Its* is used in all other cases. "*It's* bigger than *its* nest."
last / latest	The *last* is probably final; the *latest* (most recent) may be followed by others.
leave / let	Careful usage distinguishes *leave* (to depart) from *let* (to permit). "*Let* me go, or *leave* me."
luxuriant / luxurious	*Luxuriant* usually refers to lush plant growth, *luxurious* to something rich or costly. "*Luxuriant* foliage surrounded the *luxurious* summerhouse."
moral / morale	*Morale* refers to attitude or esprit de corps, *moral* to a notion of right conduct. "Our *morale* was highest in midmorning." "His *morals* are questionable."

oral / verbal	*Oral* (spoken) has a narrower meaning than *verbal* (in words, spoken or written). "He is a master of *verbal* communication, both written and *oral*."
principle / principal	(Homonyms) *Principle* (a basic doctrine or premise), *principal* (chief, main). "The *principal* speaker stood on his *principles*."
raise / rise	*Raise* is a transitive verb (to lift [something]), *rise* is intransitive (to get up). "Juliet will *rise* at dawn and *raise* the window."
set / sit	*Set* is transitive (to put something in place), *sit* is intransitive (to rest on one's haunches). "Romeo *set* up a chair and asked Juliet to *sit* in it."
stationary / stationery	(Homonyms) *Stationary* is an adjective (fixed in place), *stationery* is a noun (writing paper).
than / then	*Than* is a conjunction indicating comparison, *then* is an adverb (at that time). "*Then* he was stronger *than* he is now."

Occasionally you may be trapped into using the wrong word by some resemblance between two or three words in spelling or sound or meaning. For example, the resemblance between *refute* and *repute* led a freshman to write this sentence:

Only by knowing Communism can we *repute* it.

And the loose links of meaning between *distinguish, differentiate,* and *decipher* lured another unwary freshman to pen this statement:

Every individual, no matter what his belief may be, can *decipher* between right and wrong.

If you are uncertain of a word, look it up in your dictionary to find its exact meaning. Then an excursion into a thesaurus will give you a choice of synonyms. When you have found what you think is the right word, be sure to check it in your dictionary to make certain it is appropriate.

REVISION PRACTICE

1. Before he would except my work, I had to get an excuse from my doctor.
2. I could not run for the Dorm Counsel because of my grades.
3. Since Charles has been transferred to the missile cite, he is able to take courses at Community College.
4. My brother never liked Frank, but I find him simple and ingenious.
5. I found I could study for longer periods of time if I got up and began work at five A.M. then I could if I began at ten P.M.
6. It was an affect with their lights which they had tried for weeks to achieve.
7. He attempted to transcend to his audience all of the emotion of the poem.
8. It was the usual western with the calvary spending hours chasing Indians.

ADDITIONAL REVISION PRACTICE

1. My adviser counciled me to drop my math course and resign from the band.
2. While he said nothing, I implied from his manner that he was angry with me.
3. Since I began keeping a regular study schedule, I have had less failing grades.

4. The motor wouldn't start, and cleaning the spark plugs had no affect on it.
5. The traditional goal is to set up laws which govern mortality.
6. Since I live only sixty-three miles from here, it is very easy for me to communicate from school to home.
7. It is irrevelant to me what you do now or in the future.

3

MASTERING LARGER WRITING PROBLEMS

1

Achieving Sentence Flow

If your writing is to be clear and readable, it must have a quality that we call sentence flow. This means that each part should flow easily into the next, without gaps and without jerkiness in the forward movement of your thought. The reader should know at every point exactly where he is in relation to what has gone before. To achieve smooth sentence flow, you will find three tools of style especially useful. They are not difficult to master, and once you have learned to use them, they will do wonders in improving the readability of your writing style. These three tools are: (1) transitional devices, (2) key terms, and (3) structural links.

1. **Transitional devices** were referred to previously (*see* Section 38 T). They are words and phrases which are really not integral to the sentence but which serve to show the reader a

relationship that you wish to call to his attention. For example, the relationship of continuation is shown by terms such as *in addition, next, furthermore, besides, to continue,* and by a series of numbers. The relationship of contrast is shown by such terms as *on the contrary, however,* and *on the other hand.* The relationship of result or conclusion is indicated by such expressions as *thus, hence, so, then, consequently,* and *therefore.* They are usually placed at or near the beginning of a sentence to show the relationship of this sentence to what has gone before.

OBSERVATION EXERCISE

Underline the transitional devices used in the following sentences and indicate the relationships they show.

1. Learning to understand and use a foreign language need not be a difficult task. Moreover, in learning a new language you will also discover a new world of manners and customs and attitudes.

2. Until three decades ago, it was generally believed that there was no authentic Chinese history before the Chou dynasty, which began about 1100 B.C. Within our lifetime, however, knowledge has been so increased that we can push back the frontier of Chinese history 500 years.

3. In the Archaic Period the Greeks, believing in goodness, created sculpture that was otherworldly and unrealistic; yet, because they believed in reason, the sculpture became steadily more realistic.

4. Every name that one applies to a person is merely a label that refers to one aspect of his whole nature. For example, you can call the same person a singer, an athlete, a Negro, a lawyer, a churchgoer. Yet none of these refers to the whole person. However there is usually one name (or label or symbol) that stands out in one's consciousness above the others, though the particular name that stands out will be different in the

minds of different people. Thus the name or label we use to designate any given person distracts our attention from the concrete reality.

5. We cannot fully believe what we see, because we cannot trust our sense organs. Another person with his own eyes may see the same things we are looking at as being of a different color or shape. Besides, since his past experiences with things are different from ours, these past experiences will affect the way he perceives.

2. **Key terms** are the important, basic words that are central to what you are writing. These help you in your sentence flow when you deliberately repeat them. You may have been told in your past schooling that you should strive for variety in your word choice and should avoid repeating the same word. This advice holds true for some kinds of literary writing, but not for lucid exposition, with which we are concerned. When you are writing exposition, you must keep your meaning completely clear at all points as you advance the movement of your thought; and to keep your meaning clear, you should continue to use the same word when you refer to the same thing. For example, if you are writing about how to construct a rabbit pen, you should call it a *pen* throughout your paper and not shift to synonyms like *cage* or *hutch*. Such shifts tend to confuse the reader.

Here is an example of a careful repetition of key terms. Note how easily you can follow what this freshman is saying:

The tempo of my speech was a little fast. If I learn to slow down when I get nervous, I will have a better *speaking rate*. If my *speaking rate* is better, then my *articulation* will improve. And when I get rid of my *articulation* difficulties, people will be interested in what I have to say.

OBSERVATION EXERCISE

In the following passage the author is talking about the meanings of words. His key terms are *word* and *meaning*. Notice that he does not find it desirable to shift from *meaning* to *sense* or *signification*, nor from *word* to *term* or *expression* or *utterance*. Instead of scattering synonyms about, he repeats and repeats his key terms, thus clearing the path of thought from unnecessary obstructions. Count the repetition of *word* and *meaning*. (Italics for these words are supplied.)

> The first thing to realize is that most of the useful *words* in our language have many *meanings*. That is partly why they are so useful: they do more than double duty. Think of all the things we mean by the *word* "foot" on different occasions: one of the lower extremities of the human body, a measure of verse, the ground about a tree, twelve inches, the floor in front of the stairs, paying the bill. Yet these are pretty distinct *meanings*— they don't easily get confused with one another. It is much more difficult to distinguish clearly among the different *meanings* that the *word* "equality" takes on when we are talking about equality before the law, equality of opportunity, equality of political rights, equality of aptitude, and so forth.
>
> We can think of various *meanings* of some *words*, but we don't realize just how flexible language is until we look up some of the most changeable *words* in a large dictionary. *Webster's New International Dictionary*, for example, distinguishes twenty-four *meanings* of the *word* "free." . . . The editors of *The American College Dictionary* . . . found 55 distinct *meanings* of the *word* "point," in 1,100 occurrences of the *word*, and distinguished 109 different *meanings* of the *word* "run."
>
> *Monroe C. Beardsley*, Thinking Straight (*Prentice-Hall*)

3. **Structural links** are basic sentence parts which you place in such positions that they link the sentences together in an easy, comprehensible flow of thought. These structural links may be divided into three kinds: a. *hook-and-eye links*, b. *demonstratives*, and c. *parallel repeats*.

a. **Hook-and-eye links** are used when you refer in the beginning of one sentence to the end of the preceding sentence, to something near the end, or even to the preceding sentence as a whole. By so doing, you hook the two sentences together. Here are three examples of hook-and-eye links:

He could hardly wait to get *under the shower. There* a gushing downburst of hot, soothing water would relax his aching muscles and tired mind.

Jim formed the habit of *reviewing his eight o'clock assignment during the quarter hour just before he went to bed. By this means* he could face his first morning class with the comfortable feeling that he had the assigned reading clearly in mind.

His first step was to *clean the car thoroughly. When he had finished this job,* he was ready to apply the wax.

b. **Demonstratives** may be defined, for our purpose, as a specific set of words that point back to something which has gone before, thereby linking this something to the immediate reference. The most common demonstratives are *this, these, that, those,* and *such.* Of these five, the word *this* is the most useful and also the most treacherous. A student sentence will illustrate this treachery:

My mid-term grades were dangerously low, and my counselor advised me to spend less time on social activities. *This* made me determine to budget my time.

Here the word *this* may point back to either the first or the second part of the preceding sentence. The solution of this difficulty is simple. All you have to do is to insert a word or two after *this* specifying what *this* points back to. Our illustrative sentence would have been clear if the student had written it in either of these ways:

My mid-term grades were dangerously low, and my counselor advised me to spend less time on social activities. *This advice* made me determine to budget my time.

My mid-term grades were dangerously low, and my counselor advised me to spend less time on social activities. *This low standing* made me determine to budget my time.

To see further how demonstratives operate as structural links, you might reread the third and fourth sentences of this paragraph.

c. **Parallel repeats** are repetitions of a grammatical pattern (*see* Section 23 Paral). Such repetitions may occur within a single sentence, in a series of sentences, and even in a sequence of paragraphs. The next four examples illustrate parallel repeats within sentences:

1. Studies serve *for delight, for ornament,* and *for ability.* Their chief use *for delight is in privateness and retiring; for ornament, is in discourse;* and *for ability, is in the judgment and disposition of business.*

Francis Bacon, "Of Studies"

2. It was not so much by a *knowledge of words that I came to the understanding of things* as *by my experience of things that I came to follow the meaning of words.*

Plutarch

3. Vigorous writing is concise. *A sentence* should contain *no unnecessary* words, *a paragraph no unnecessary* sentences, for the same reason that *a drawing* should have *no unnecessary* lines and *a machine no unnecessary* parts. This requires *not that* the writer make all his sentences short, *or that* he avoid all detail and treat his subjects only in outline, *but that* every word tell.

William Strunk, Jr., and E. B. White, Elements of Style (*Macmillan*)

4. Thought marches; and marches to the rhythm of subject and predicate. *The foot that is planted on the ground represents the subject,* and *the foot that is moving forward in the air is the predicate. All the stability is in the subject, all the movement in the predicate.*

P. B. Ballard, Thought and Language (*University of London Press*)

In the following example parallel repeats assist in directing the traffic of thought within the long sentences and in the successive short ones:

For instance, *let a person, whose* experience has hitherto been confined to the more calm and unpretending scenery of these islands, *whether* here or in England, *go* for the first time into parts where physical nature puts on her wilder and more awful forms, *whether* at home or abroad, as into mountainous districts; or *let one, who* has ever lived in a quiet village, *go* for the first time to a great metropolis,—then I suppose he will have a sensation which perhaps he never had before. *He has* a feeling not in addition or increase of former feelings, but of something different in its nature. *He will* perhaps be borne forward, and find for a time that he has lost his bearings. *He has* made a certain progress, and *he has* a consciousness of mental enlargement; *he does* not stand where he did, *he has* a new centre, and a range of thoughts to which he was before a stranger.

John Henry Newman, The Idea of a University (*Longmans*)

In the following example parallel repeats assist in linking sentences together:

If we wish to play an effective part as members of a community, we must avoid two opposed dangers. *On the one hand there is the danger of rushing* into action without thinking about what we are doing. *On the other hand there is the danger of indulging* in an academic detachment from life. This is the peculiar temptation of those who are prone to see both sides of a question and are content to enjoy an argument for its own sake. But thinking is primarily for the sake of action. *No one can avoid* the responsibility of acting in accordance with his mode of thinking. *No one can act* wisely who has never felt the need to pause to think about how he is going to act and why he decides to act as he does.

L. Susan Stebbing, Thinking to Some Purpose (*Penguin*)

The next selection shows us how parallel repeats are used to help link a sequence of paragraphs together:

The politicians tell us, "you must educate the masses because they are going to be masters." *The clergy join in* the cry for education, for they affirm that the people are drifting away from the church and chapel into the broadest infidelity. *The manufacturers and the capitalists swell the chorus lustily.* They declare that ignorance makes bad workmen. . . . *And a few voices are lifted up* in favor of the doctrine that the masses should be educated because they are men and women with unlimited capacities for being, doing, and suffering, and that it is as true now, as ever it was, that the people perish for lack of knowledge.

These members of the minority, with whom I confess I have a good deal of sympathy, *question* if it be wise to tell people that you will do for them, out of the fear of their power, what you have left undone, so long as your only motive was compassion for their weakness and their sorrows. And if ignorance of everything which it is needful a ruler should know is likely to do so much harm in the governing classes of the future, why is it, they ask reasonably enough, that such ignorance in the governing classes of the past had not been viewed with equal horror? . . .

Again, *this sceptical minority asks* the clergy to think whether it is really want of education which keeps the masses away from their ministrations—whether the most completely educated men are not as open to reproach on this score as the workmen; and whether, perchance, this may not indicate that it is not education which lies at the bottom of this matter?

Once more, *those people,* whom there is no pleasing, *venture to doubt* whether the glory, which rests upon being able to undersell all the rest of the world, is a very safe kind of glory —whether we may not purchase it too dear; especially if we allow education, which ought to be directed to the making of men, to be diverted into a process of manufacturing human tools, wonderfully adroit in the exercise of some technical industry, but good for nothing else.

And finally, *these people inquire* whether it is the masses alone who need a reformed and improved education. . . . They seem to think that the noble foundations of our old universities are hardly fulfilling their functions. . . . And while as zealous for education as the rest, they affirm that if the education of the richer classes were such as to fit them to be the leaders and the governors of the poorer; and if the education of the poorer classes were such as to enable them to appreciate really wise guidance and good governance, the politicians need not fear mob-law, nor the clergy lament their want of flocks, nor the capitalists prognosticate the annihilation of the prosperity of the country.

Thomas H. Huxley, Collected Essays (*Houghton Mifflin*)

We have now discussed five tools of style: transitional devices, key terms, and three structural links—hook-and-eye links, demonstratives, and parallel repeats. If you learn to employ these skillfully, you will be able to avoid the rough, jouncing style common to beginning writers and to achieve a sentence flow that will make your writing easy to follow and understand.

OBSERVATION EXERCISE

The passages below are the work of seasoned writers. Study them carefully for sentence flow, underlining the instances you find of transitional devices, repetition of key terms, hook-and-eye links, demonstratives, and parallel repeats.

1. A speaker commands resources of expression far richer than those of a writer. He can reinforce particular points by giving special emphasis of voice and intonation to them; he can make use of facial gesture. He can play tricks with his hands and fingers, opening and shutting them, waving them up and down and sideways. If he is near enough to his victim, he can even nudge him to drive home a specially important sally, although modern ideas of good manners tend to look on this practice as low-bred.

Adapted from *Hugh Sykes Davies*, Grammar Without Tears (*John Day*)

2. Fifty years ago people in this country believed that "love apples" (tomatoes) were poisonous, because everybody believed that love apples were poisonous. This superstition, like many others, died a natural death without causing any serious hardship. But when potatoes were first introduced in Russia,

the peasants would not plant or eat them, because, they said, they were "devil's apples." The rulers realized that the potato could become an important factor in agricultural economy; so they tried to overcome the peasants' reluctance by *forcing* them to eat potatoes. The question how to verify the assertion "potatoes are devil's apples" never arose. The peasants said it was "true," and the officials said it was "false." This controversy caused a great deal of unhappiness.

Anatol Rapoport, Science and the Goals of Man (*Harper*)

3. How does it happen that children in general learn their mother-tongue so well? That this is a problem becomes clear when we contrast a child's first acquisition of its mother-tongue with the later acquisition of any foreign tongue. The contrast is indeed striking and manifold: here we have a quite little child, without experience or prepossessions; there a bigger child, or it may be a grown-up person with all sorts of knowledge and powers: here a haphazard method of procedure; there the whole task laid out in a system . . . : here no professional teachers, but chance parents, brothers and sisters, nursery-mates and playmates; there teachers trained for many years specially to teach languages: here only oral instruction; there not only that, but reading-books, dictionaries, and other assistance. And yet this is the result: here complete and exact command of the language as a native speaks it, however stupid the children; there, in most cases, even with people otherwise highly gifted, a defective and inexact command of language. On what does this difference depend?

Otto Jespersen, Language: Its Nature, Development and Origin (*Macmillan*)

4. Man is a rational animal—so at least I have been told. Throughout a long life, I have looked diligently for evidence in favor of this statement, but so far I have not had the good fortune to come across it, though I have searched in many countries spread over three continents. On the contrary, I have seen the world plunging continually further into madness. I have seen great nations, formerly leaders of civilization, led astray by preachers of bombastic nonsense. I have seen cruelty, per-

secution, and superstition increasing by leaps and bounds, until we have almost reached the point where praise of rationality is held to mark a man as an old fogey regrettably surviving from a bygone age. All this is depressing, but gloom is a useless emotion. In order to escape from it, I have been driven to study the past with more attention than I had formerly given to it, and have found, as Erasmus found, that folly is perennial and yet the human race has survived. The follies of our times are easier to bear when they are seen against a background of past follies.

Bertrand Russell, "An Outline of Intellectual Rubbish," Unpopular Essays (*Simon and Schuster*)

5. Some forms of ineffective thinking are due to our not unnatural desire to have confident beliefs about complicated matters with regard to which we must take some action or other. We are sometimes too lazy, usually too busy, and often too ignorant to think out what is involved in the statements we so readily accept. Few true statements about a complicated state of affairs can be expressed in a single sentence. Our need to have definite beliefs to hold on to is great; the difficulty in mastering the evidence upon which such beliefs ought to be based is burdensome; consequently, we easily fall into the habit of accepting compressed statements which save us from the trouble of thinking. Thus arises what I shall call 'Potted Thinking.'

L. Susan Stebbing, Thinking to Some Purpose (*Penguin*)

6. Language is nothing but a set of human habits, the purpose of which is to give expression to thoughts and feelings, and especially to impart them to others. As with other habits it is not to be expected that they should be perfectly consistent. No one can speak exactly the same as everybody else or speak exactly in the same way under all circumstances and at all moments, hence a good deal of vacillation here and there. The divergencies would certainly be greater if it were not for the fact that the chief purpose of language is to make oneself understood by other members of the same community; this presupposes and brings about a more or less complete agree-

ment on all essential points. The closer and more intimate the social life of a community is, the greater will be the concordance in speech between its members. In old times, when communication between various parts of the country was not easy and when the population was, on the whole, very stationary, a great many local dialects arose which differed very considerably from one another. . . . In recent times the enormously increased facilities of communication have to a great extent counteracted the tendency towards the splitting up of the language into dialects. . . .

Otto Jespersen, Essentials of English Grammar (*Holt, Rinehart and Winston*)

7. Darwin's hypothesis as contained in his *Origin of Species* (1859) is known as the hypothesis of natural selection. This involves the idea that it is nature, or the environment, which selects those variants among the offspring that are to survive and reproduce. Darwin pointed out, first of all, that the parents of every species beget more offspring than can possibly survive. He maintained that, consequently, a struggle takes place among these offspring for food, shelter, warmth, and other conditions necessary for life. In this struggle certain individuals have the advantage because of the factor of *variation*, which means that no two of the offspring are exactly alike. Some are born strong, others weak; some have longer horns or sharper claws than their brothers and sisters or perhaps a coloration of body which enables them better to blend with their surroundings and thus to elude their enemies. It is these favored members of the species that win out in the struggle for existence; the others are eliminated generally before they have lived long enough to reproduce.

Edward M. Burns, Western Civilizations, *3rd ed.* (*Norton*)

REVISION PRACTICE

Rewrite the following sentences to achieve a smoother flow of thought. In each case use the devices indicated in parentheses.

1. In ancient Babylon much business was carried on near

the gate of the city. A crowd of people chatting together and waiting for news from travelers could always be found there. (hook-and-eye link)

2. Variety of pitch is another voice quality that is lacking in my speech delivery. The absence of this factor tends to make my speeches monotonous. (key term)

3. In the course of college life one is constantly making new friends. One can hear of a rich variety of experiences from these new acquaintances during dormitory talk fests. (key term and hook-and-eye link)

4. The courses in the general education program give a student a broad and systematic basic knowledge. For example, in biological science he learns of life processes from the lowest forms up to men. He becomes acquainted with the music of the great composers of our western tradition in the course called exploring music. The two courses in humanities teach him the history of western Europe and acquaint him with important social and literary documents. (parallel repeats)

5. An excellent way to learn to understand French is to listen to it in the language laboratory. A student hears the language spoken easily and naturally by a native by means of the tapes there. He can listen by himself and take all the time he wants to master his lesson. (hook-and-eye and transitional device)

6. One of the earlier Italian masters of painting and sculpture was Andrea Verrocchio. Leonardo da Vinci first learned the craft of painting from him. (hook-and-eye link)

7. In the North of England the traveler passes through the wild green mountains of Wordsworth's Lake District and up to the lowlands of Burns. He sees the "checkered counties" of Housman in the West. The South reveals to him the heaths and moors made famous by Hardy. (parallel repeats)

8. There are three things you should do in the introduction of your speech. You should capture the attention

of your audience. This can be done in various ways, such as making a startling statement, or telling a personal story, or asking the audience a question. You should create a favorable impression toward yourself. This you can do by being natural and dignified and by treating your listeners with respect. You should lead into the body of your speech. A simple way to do this is to state your purpose. (transitional devices)

9. This paper is divided into two sections for two reasons. First, it gives the reader some facts and figures on the advancements made in the field of jet aviation. Second, the last section of the paper was written for the purpose of informing the reader when and where the first jet aircraft was invented. (parallel repeats)

10. I had put off writing my term paper until the day before it was due. I had to sit up nearly all night to get it written in time for my eight o'clock class. (transitional device)

ADDITIONAL REVISION PRACTICE

1. The by-products of man's attempt to master science have been numerous. Our society has leisure time and industrial development never before experienced. (Repeat key term "One by-product. . . .")

2. A good lecturer can hold the attention of his students. It is necessary to give vocal emphasis to important statements. Humor helps to break up the monotony of a class lecture. (Try two ways here. (1) Make passage into one sentence using parallel repeats in -ing. (2) Use hook-and-eye at beginning of second sentence —"To do"—and put last two sentences into one, using parallel repeats.)

3. The worst part about writing an in-class theme is getting organized. Some people get poor grades because they cannot organize their ideas. Students often give up in despair. (Use hook-and-eye at beginning of sec-

ond sentence and combine the last two into one. Avoid shift from *people* to *students*.)

4. Survival is not merely a matter of keeping body and soul together, as it was in the jungle. You must keep within the good graces of your superiors. You must behave properly, which means putting up a courteous and hospitable front at all times. (Use parallel repeat "It is a matter. . . .")

5a. Lincoln Highway is a road that runs east and west along the edge of the campus. North of Lincoln Highway, where it meets Castle Drive, are the Montgomery Arboretum and the Lagoon. There are houses on the south side. (Make change only in third sentence.)

5b. Lincoln Highway is like a broad river running along the campus and through the town. The banks of the river change color in the fall. The arboretum changes to yellow and orange as the frost kills the leaves. The leaves fall from the trees and dye the road a dirty yellow. (Use hook-and-eye at beginning of third sentence, and combine last two sentences into one.)

2

Developing Thought in Paragraphs

A paragraph is a block of print or writing that usually begins with an indentation and is built around a single topic, or a single aspect of a topic. You will do wisely in your themes to follow this principle. Thus each paragraph indentation will tell your reader that he has come to another topic, another aspect of the topic, or another division in thought.

Paragraphs vary in length from one word or phrase to hundreds of lines, and you may well wonder, "How long should I make my paragraphs?" A good answer lies at hand in your college textbooks of history, science, humanities, and the like. Open a few of them and see for yourself how long the paragraphs are. In normal expository prose you will probably find from one to three paragraph indentations on each printed page. Then a little word-counting and averaging will give you a rough clue to normal paragraph length.

But your main guide is your topic. A rather short topic might be dealt with in a single longish paragraph, whereas a longer topic may naturally break into several divisions, each deserving a paragraph by itself. Many freshmen, beginning their composition course, fall into the error of using too many paragraphs. If you look at your theme and see a paragraph indentation every two or three lines, something is probably wrong. Either these paragraphs are not fully developed, or they should be regrouped into longer paragraphs.

A fully developed paragraph must say enough to clarify or explain its topic for your reader. The undeveloped paragraph leaves its generalization unsupported, unexplained. Here is an example of an undeveloped paragraph:

> One might ask *why I did not go to college in my home town.* To answer this question is somewhat difficult, although I feel my reason is substantial. I know several graduates of City College, and few of them seem to have gotten anything out of their education.

This paragraph raises a question and leaves it in the air. The second sentence is a vague generalization, and the last sentence fails to give the "substantial reason" which the writer has promised.

Filling out paragraphs requires a good deal of practice in thought development. A helpful preliminary to this practice is reading some well-written paragraphs and tracing their patterns of thought development. In the following pages you will be given examples of some of the more useful patterns, which you may find profitable to imitate as you develop the various topics that confront you in your writing. The patterns are often given these names:

1. Example
2. Time arrangement
3. Space arrangement
4. General-and-specific
5. Statistics
6. Comparison-contrast
7. Division or classification
8. Cause-and-effect
9. Analogy

To this list we must append several qualifications. First, there are more patterns than those listed. Second, the categories overlap. That is to say, you will meet with passages which may be classified according to two or more of the labels on the list. Third, you will come across sound and logical writing that does not seem to follow any identifiable pattern of development and yet is clear and cogent. Nevertheless, an acquaintance with these patterns will be of value to you as you set about the arduous task of learning to write clear and simple prose.

SAMPLES OF THOUGHT DEVELOPMENT

1. Development by Example

This is a very common and useful method of thought development. It is quite simple. The writer makes a statement and then uses one or more examples to illustrate, support, and make clear the statement. The order is sometimes reversed, proceeding from the example to the statement they illustrate.

> TOPIC: *Compound superlatives in the language of the Ozarks.*
> DEVELOPMENT: *Numerous examples.*

a. There are many compound superlatives in the language of the Ozarks, such as *loud-cussin'est, hell-raisin'est, fish-ketchin'- est, vote-gettin'est,* and *rabbit-killin'est.* A man once told me, "Katy is the most *out-doin'est* woman that ever lived," and I think he meant that she was surprisingly vigorous and energetic. One of my friends was badly cut in a knife fight and required many stitches to repair his injuries; of the physician who did the work he remarked, "Doc Holton is the *stitch- takin'est* feller I ever met up with." A boatman who prepared meals for tourists on a float-trip was described as "the *pancake- cookin'est* feller on the creek." A man who sold out several times and moved to Oklahoma, invariably returning a few months later, was referred to as "the *back-comin'est* feller in this country." An editorial in a Little Rock newspaper points out that the United States is "the *statistics-keepingest* nation on the face of the earth." A certain senator in southwest Missouri was described as "the *potguttedest* candidate that ever crawled up on a stump." A mountain man remarked to me that his children were maturing very rapidly, adding, "I believe Lolly is the *growed-uppest* one of the lot." Such superlatives as *sleepy-headedest* and *high-poweredest* are familiar to everybody in the back hills.

From Vance Randolph and George P. Wilson, Down in the Holler, (*University of Oklahoma Press*)

TOPIC: *Scientific generalizations are subject to change.*
DEVELOPMENT: *Use of a single hypothetical example.*

b. A scientific generalization is always subject to change in the light of further evidence. As an example, let us see what could happen to Cuvier's famous generalization that all animals with both horns and hoofs eat grasses and grains, not flesh. Suppose that a scientist in South America has discovered a large plateau high among the mountains of the Andes. As he and his party begin to cross this plateau they notice in the distance a herd of strange animals which seem to be grazing. Approaching more closely, they take out their binoculars and inspect the herd carefully. They note that these animals all have horns and hoofs. But they also discover, to their amazement, that instead of grazing, some of the animals are eating rabbits, which they have apparently killed. This evidence makes it necessary to change the generalization they had held—that *all* animals with horns and hoofs are granivorous, not carnivorous.

Practice topics for development by examples:

1. I have learned a lot about study methods in the last month.
2. Haste makes waste.
3. Sports in high school are good for a student.
4. My home town has much to be proud of.
5. Debating can be a valuable addition to a student's education.
6. Blind dating is hazardous.
7. Owning a car is detrimental to a student's academic record.
8. I have been surprised to see how, as I grow older, my father seems to get smarter.

2. Development by Time Arrangement

When we tell a sequence of events in the order of their occurrence, we are using development by time. We frequently use this method in relating experiences and spinning yarns. Time order can be juggled around. For instance, you can begin with an exciting moment and then go back and show what led up to this moment.

TOPIC: *History of the word* nice.
DEVELOPMENT: *Changes in the meaning of* nice *during the passage of centuries.*

The word *nice*, which today is a verbal factotum to indicate approval, has had a long and interesting history. We first meet it in Roman times as the Latin adjective *nescius,* which meant "not knowing" or "ignorant." Then, when the Roman armies invaded what is now France and when the Roman traders and settlers followed them, they naturally brought with them their native Latin tongue, including the word *nescius.* The speech of Rome became the speech of Gaul (France). In the course of centuries *nescius* changed in form and sound, and when we next meet it—in the twelfth century in the work of Chrestien de Troyes—it had become *nice,* pronounced like "niece." It had also taken on another meaning, that of "foolish." Next, the word *nice* entered England. We all remember the Norman Conquest of 1066, when the Norman French conquered England, and the period of French domination that followed. During these centuries the French and English languages existed side by side, and it was inevitable that many French words should become naturalized in English. One of these was *nice,* and it first appears in written records as an English word in the fourteenth century. Its meaning was "foolish," and it was regularly used in this sense by Chaucer, the great fourteenth century poet. As time went on, its meaning became narrowed to "foolishly particular about small things." From this meaning it was an easy step to the next one, "particular about small things," and hence "discriminating," "accurate." Thus one could speak of a "nice observer of human foibles" or a "nice taste in wine" or a "nice distinction." The final change in meaning was from "discriminating" to "agreeable" or "excellent." What a nice (discriminating) person with a nice (accurate) taste would choose would be a nice (excellent) thing. This change appeared in the eighteenth century. The last two meanings of *nice* are in use today. Careful writers and speakers often use it to mean "discriminating," and everyone uses it in ordinary speech in the general sense of "agreeable" and "pleasant."

Practice topics for development based on time sequence:

1. The last days before graduation stand out clearly in my memory.
2. We took an unusual trip last summer.
3. Recently we had to pack up and move into a new house, all in one week.
4. A basketball player must keep to a strict schedule when preparing for the season.
5. You cannot imagine the trouble I had getting a book out of the library.

6. It took three weeks to plan our vacation trip.

7. You must begin preparing for a party several days or a pencil, or a dime.

3. Development by Space Arrangement

Here you take your reader from one place or position to another in an orderly fashion. You would use this method if you were to describe such things as a college campus, or your room, or a pencil, or a dime.

> TOPIC: *English words have different meanings in different parts of the world.*
>
> DEVELOPMENT: *Details are arranged in the spatial order of England, Australia, South Africa, and the United States.*

    The words and expressions that constitute the word stock of the English language differ considerably in different parts of the English-speaking world. In England, for example, the motor of a car is under a *bonnet*, not a *hood*, and the riders in the front seat are protected from the wind by a *windscreen*. An Englishman will do *straightaway* what an American will do *right away*. One rides the *underground* in London, not the *subway*. A well-dressed man may be wearing a *bowler* (derby), and he holds up his socks not his trousers with his *suspenders*. In Australia one hears the term *dinky-dy* in place of our *OK*. What is a *ranch* to us is a *station* to an Australian. The character that we know as a *hobo* or *tramp* is known in the continent down under as a *swagman* or *swaggie*. He travels through the *outback* or *bush* (back country) and carries his *swag* (bundle). In South Africa English has acquired many new words, especially from the local variety of Dutch called Afrikaans. A village is known as a *dorp*. An overseer or foreman is a *bass*, a Dutch word which was independently adopted into American English in the form of *boss*. A hill is a *kopje*, a word that we meet in John Masefield's famous poem, "A Consecration," in the form of *koppie*. A movie is a *bioscope*, a wagon trip is a *trek*, and a porch is a *stoep*, which also appears in the United States as *stoop*. In the United States we find local differences in the word stock. In Boston, for instance, one drinks *tonic,* whereas in Chicago the same beverage is *pop.* When a woman takes her husband to work in the family car, she *carries* him to work in Texas and *drives* him to work in Iowa. A *fried cake* in upper New York state is a *cruller* in New Jersey and a *doughnut* in Minneapolis. The plural of *you* is *you-all* throughout the South but in western Pennsylvania it becomes *you'ns.* We see, then, that the vocabulary of the English language is not uniform but is different wherever we go.

Practice topics for development by spatial arrangement:

1. The new Mustang has an unusual instrument panel.
2. My Father's workshop shows how meticulous he is.
3. My room is arranged to suit my special needs.
4. When I can afford it, I'm going to have a perfect stereo system.
5. Our campus is easy to get around, once you understand its layout.
6. Our college library is well planned for browsing.
7. My state has a wide variety of scenic interests.
8. New York is easy to get around in.

4. Development by General-and-Specific

In this method, you may begin with a general statement and continue by presenting specific details that bear out, expand, and support the general statement. This order is illustrated in the first example below. Or you may reverse this order, mentioning the specifics first and then tying them together with the generalization at the end. This reversed order is shown in the second example.

TOPIC: *The American office as a temple of status.*
DEVELOPMENT: *Listing of specific details.*

a. The American office is a veritable temple of status. Though they may seem almost imperceptible, the symbols are manifested everywhere. Some have a useful purpose—the memo pad "From the desk of . . ."; the routing slip (Should the names

on the memorandum be listed by seniority or alphabetically?);
who sits with whom in the company dining room. Others are
rooted in propriety; who can call whom by nickname, at what
level may people smoke? To what grade of washroom is one
entitled? Is the office carpeted or does he rate only linoleum?
Some are rooted in functions only marginal: the facsimile
signature stamp, for example—evidence that a man's importance is such that he must write to a great number of people,
even if he doesn't use the facsimile signature in doing it. All
these are favorite topics of office humor, of course, but as the
fact itself is witness, the symbols communicate.

Adapted from William H. Whyte, Jr., Is Anybody Listening? (*Simon
and Schuster*)

TOPIC: *The effectiveness of written style as compared
            with colloquial style.*
DEVELOPMENT: *Description of an experiment—its method,
            conditions, subjects—and a resultant general-
            ization.*

b. A British professor recently conducted a revealing experiment in language. He constructed a short message and arranged
the words in two different ways. One was an arrangement characteristic of written style; it went like this:

He's doing research on the procedures for assessing, the methods of surveying, and the techniques for exploiting the mineral
resources of the various parts of the Commonwealth.

The second arrangement was a colloquial one, that is, characteristic of spoken style. It contained the same information
and the same words:

He's doing research on the mineral resources of the various
parts of the Commonwealth—the procedures for assessing, the
methods of surveying, and the techniques for exploiting them.

These two arrangements were tested out on two groups of
undergraduates, who had no idea of what was involved. Each
group consisted of about thirty-five first-and-second-year students, in roughly equal proportions. The two forms of the test
piece occupied the same reading time, and the reading rate was
very fast, as a final check against one hundred percent per-
ception and recall that would, of course, give data not sus-

ceptible of comparison. The group that received the written style absorbed on average *forty-five* percent of the information. But the group which received the colloquial style absorbed on average *sixty* percent of the information. It might seem, then, if we can generalize from so simple a case and from so few numbers, that the arrangement of words which a speaker uses can influence the retention of what he says.

Adapted from Randolph Quirk, "Colloquial English and Communication," in Studies in Communication *(Oxford University Press)*

Practice topics to be developed by proceeding from general to specific or specific to general:

1. You can recognize a New Yorker anywhere.
2. A used car is expensive for a student.
3. The assignments in this course could be improved.
4. Love conquers all.
5. A college student has a busy life.
6. Being in college brings new responsibilities.

5. Development by Statistics

In this method a conclusion is supported by figures and statistics. This method can also bear other labels, such as development from general to specific or from specific to general, depending on the position of the conclusion.

TOPIC: *English is an important language.*
DEVELOPMENT: *Statistics on the number of speakers of English 400 years ago and in our century.*

English is an important language. Only 400 years ago it was spoken by fewer than 5 million persons living on a small island off the coast of Europe. Today it spans the globe. On the original island it is now the tongue of some 40 million people. In Canada it is spoken by 11 million. In the United States an estimated 175 million use English as their daily language. Away down under in Australia, over 7 million persons have their cockney brand of English. To these figures we must add the

Englishmen scattered over the earth, the Irish, and the many Europeans to whom English is a second tongue. Thus we see that there are probably over 250 million speakers of English. And when we add to this the fact that it is the language of the most powerful and technologically advanced country in the world, we can harbor no doubt that English is an important language.

Practice topics for development by use of statistics:

1. Government aid to students is increasing year by year.
2. The average age at which couples marry is going up.
3. The increase in our standard of living may be measured by the number of luxury items owned by an average family.
4. High speed is the cause of many highway accidents.
5. Farming areas are rapidly losing their population.
6. Alcoholism has increased dramatically.
7. We are making progress in Civil Rights.

6. Development by Comparison-Contrast

You can develop a comparison by pointing out similar features. You could compare the United States and Russia in terms of size, climate, political influence, economic power. You can develop a contrast by pointing out differences between things somewhat similar. You would contrast the United States and Russia, or California and Rhode Island, but certainly not a goldfish and an elephant. Both comparison and contrast might be used in the same theme.

In general, there are two methods of developing a comparison. The first is to describe one side of the comparison as fully as you intend and then deal entirely with the other side. When you practice this method be sure to follow the same order of points of comparison for each side. The first example which follows shows this method. When you use the first method, be sure your transitions are clear. The second compares the two activities point by point.

TOPIC: *Differences between high school and college life.*

DEVELOPMENT: *The details of high school life are all given first; then a description of college life is presented.*

a. High school life and college life have startling differences. In high school my teachers were always harassing me about the quality of my work and about late or unfinished papers. I seemed to have all the time I needed for extracurricular activities. My mother's good meals, my own quiet room, and all my pals at East High I took for granted. In college, my life suddenly became different. The responsibility for quality work and for promptness in handing in my papers was left strictly up to me. The first college year has left me little time for outside activities, and I have had to budget my time carefully. The cafeteria meals, my noisy room, and my new friends are all a contrast to my previous life. Thoughts of home life sometimes keep me awake at night after the clamor of the dormitory has subsided. However, each new day with its challenges, excitement, and independence brings stimulation that makes memories of high school life seem dull indeed.

TOPIC: *Reading and listening a means of learning.*

DEVELOPMENT: *Point-by-point comparison.*

b. The two communicative skills of reading and listening, research tells us, are about equally effective as means of learning. But from the point of view of the receptor, reading is sometimes better. *Reading* has the advantage of being much *faster* than listening. An average normal rate for an adult reader is perhaps 300 words a minute, although many can exceed this by a good deal. In listening, however, one's rate is held down to that of the speaker, from about 125 or 175 words a minute. In *reading* we can *set our own pace*, taking it fast or slowly as we wish, and when we're tired we can take a break. But in listening we must take the pace set for us by the speaker. If it is too slow for our condition of alertness, we may tend to daydream; and if it is too fast for the difficulty of the material, we may become fatigued and lose the thread of thought. When we are *reading we can* always *stop* to reread a difficult passage, and we have all the time we want to look up things that will help us understand what we are reading. Listening, however, is another story. We must catch everything on the fly or else it is lost. If our attention wanders even for a few seconds, we may miss an important sentence that is a key to what follows.

Practice topics for development by comparison-contrast:

1. A doctor is more valuable to the community than a lawyer.
2. A Mustang is a better buy than a Corvette.
3. Traveling by train is more fun than flying.
4. Regular study brings better results than periodic cramming.
5. Basketball is a more demanding game than tennis.
6. Our popular music is in some ways like the music of the Roaring Twenties.
7. My math course makes different demands on me than does my English course.
8. The ability to speak persuasively is worth more than the ability to reason.
9. The library is a better place to study than the dorm.
10. The Midwest is more friendly than the East.

7. Development by Division or Classification

In using this method, you first divide the topic or main point into several parts. Then you develop each part in turn, using any of the methods we are dealing with here. In the selection that follows, notice that the topic is divided into four parts, each of which is developed by examples.

TOPIC: *Making new words.*
DEVELOPMENT: *Dividing the material into four groups. In a long paragraph like the foregoing, it is sometimes helpful to number the divisions of the topic.*

In English there are four ways in which new words can be made: by composition, by derivation, by sound-symbolism, and by root-juncture. The first one, composition, means the joining together of two existing words to form a compound word. From this process we have got thousands of compound words, such as *cherry tree, blackbird, dark-red, breakfast, overcome.* In the most frequent kind of compound word the last part of the word has a general meaning which is made more specific by the first part, as in *race horse, horse race, dog house,* and *house dog.* Other kinds of compound words are so many and the relationships between their parts so complex that it would not be profitable to explore the subject in our limited space. Derivation means making a new word by adding a meaningful prefix or suffix to an already existing word. As examples of words made by the addition of a prefix, we may cite *intramural, counteract, impossible, coexist, intercollegiate.* Words that have been made by the addition of a suffix may be exemplified by *hoggish, goodness, draftee, waiter, waitress, cigarette, lemonade, noisy,* and *sweetly.* The number of meaningful suffixes available in English for derivation greatly exceeds the number of prefixes. Sound-symbolism means the inventing of a new word in which the sounds in it resemble its meaning. Words like *sizzle, ping pong, bang, roar, thump, see-saw, bobwhite, slurp* are common examples of sound symbolism. The process is an active one, as are the two mentioned above, and the observant student will find many instances if he listens carefully to the language around him, especially slang. Root-juncture means the formation of a new word by joining together two roots, usually from the Latin or Greek. *Biology,* for example, is formed from two Greek-derived roots—*bio,* meaning life, and -*logy,* meaning a theory or a science or a doctrine. Thus biology is the science of life. Other examples of root-junctures are *audiophile, telephone, anthropology, phonograph, gramophone, circumspect, introvert.* Today this method of forming new words is used largely in science.

Practice topics for development by division or classification:

1. The animals which make the best pets for children are dogs, rabbits, and squirrels.
2. After attending classes at college for several weeks, I have discovered four sure ways of identifying a dull teacher.

3. While working in a library, I learned that there are four classes of library visitors: those who come to look for a date, those who come to read the magazines, those who rush in to cram for an hour at the reserve shelf before a quiz, and those who are interested only in taking out books.

4. A visitor to my home town will find at least three kinds of entertainment.

5. In my dormitory there are three kinds of students who, I believe, are not getting the most out of college. These are the Greasy Grind, the Social Butterfly, and the Activities Chaser.

6. Sitting near me in class are several students who have found foolproof ways to make their college life miserable: the Hair-Shirt, the hand-holding Lovers, and the Sleeper.

8. Development by Cause and Effect

Here you have a choice of two procedures. The first is to describe a situation and then to show what has caused it. This is done in the first illustrative paragraph. The second is to deal with the causes first and then to present the effects or results stemming from these causes. The second example shows this procedure.

TOPIC: *The problems the college freshman faces in learning to speak before a group.*

DEVELOPMENT: *The problems of the freshman speaker are listed and the causes are described.*

a. The college freshman who is learning to speak before a group is sometimes ineffective at first and is unable to convey to the group exactly the message or point that he is trying to make. And he is frequently surprised to find out that his listeners did not grasp the point, for it seemed perfectly clear to him. This ineffectiveness is usually caused by three difficulties, each of which can be overcome with practice and care. The first is that the speaker does not give extra emphasis to the important parts of his talk. Instead, everything that he is putting forth rolls along flatly and without variation; there are no hills and plains. The consequence of this unrelieved sameness is that listeners have trouble in separating the important from the unimportant, the significant from the trivial; and at the end they are unsure of the point. The second difficulty is that the speaker does not set up signposts along the way. He forgets, for instance, to number his points. He forgets to pause between points. He forgets to use relational expressions to show thought relationships—such expressions as *in the first place, as a beginning; on the other hand, on the contrary, nevertheless; in addition, to continue, next; hence, thus, consequently; for example, for instance, to illustrate; to conclude, in conclusion, in short, finally, to summarize, then.* Without such signposts listeners sometimes get lost, and then they lose interest and begin to think about other things. The final difficulty is that the beginner's rate of speed is far too rapid. In his nervousness and in his desire to get along with his subject, he speeds ahead at a pace that makes listening difficult. And when listening becomes difficult, listeners tend to give up. These three difficulties are the cause of much of the ineffectiveness of talks given by beginning speakers before college groups.

TOPIC: *Dangers freshman students face in beginning college work.*

DEVELOPMENT: *Some of the causes of freshmen difficulties are discussed, leading to an obvious result.*

b. The dangers faced by freshman students as they begin their college work can have disastrous results. If they have not learned to study effectively, they may spend hours on what should be a reasonably short reading assignment. They may fail to budget their time and discover that the day is too short for all they must cram into it. They may let themselves be enticed into social programs and college activities to the extent that there is insufficient time left for the demanding work of the classroom and laboratory. The result is that, at the end of the semester, they receive a "Dear John" letter from the dean indicating that they must leave college to make room for more promising students.

Practice topics for development by cause-to-effect or effect-to-cause:

1. Dancing is popular among young people.
2. There is a great need for public, noncommercial television.
3. Comic books (or the Beatles, or psychedelic shops) are popular today.
4. Beards are generally disapproved of in our society.
5. What are the common causes of academic failure among college freshmen?
6. A poor examination grade may be explained in several ways.
7. Failure to obey traffic laws can be dangerous.
8. A college student should learn to type.

9. Development by Analogy

An analogy is a special kind of comparison. The items compared are usually things that one considers quite unlike in most respects, such as an automobile engine and the human body, a garden and a college, a house and a book. An analogy often proceeds, point by point, for considerable length.

TOPIC: *The structure of a book.*

DEVELOPMENT: *Similarities between the architecture of a house and the structure of a book are discussed.*

A book is like a single house. It is a mansion of many rooms —rooms on different levels, of different sizes and shapes, with different outlooks, rooms with different functions to perform. These rooms are independent, in part. Each has its own structure and interior decoration. But they are not absolutely independent and separate. They are connected by doors and arches, by corridors and stairways. Because they are connected, the partial function which each performs contributes its share to the usefulness of the whole house. Otherwise the house would not be genuinely livable.

The architectural analogy is almost perfect. A good book, like a good house, is an orderly arrangement of parts. Each major part has a certain amount of independence. As we shall see, it may have an interior structure of its own. But it must also be connected with the other parts—that is, related to them functionally—for otherwise it could not contribute its share to the intelligibility of the whole.

As houses are more or less livable, so books are more or less readable. The most readable book is an architectural achievement on the part of the author. The best books are those that have the most intelligible structure and, I might add, the most apparent.

From *Mortimer Adler*, How to Read a Book (*Simon and Schuster*)

Practice topics for development by analogy:

1. The human life cycle is like the four seasons.
2. The human heart is like a fuel pump.
3. A college is like a supermarket (or an automobile-production plant).
4. College life is like a window-shopping tour (a track-meet, a bridge, a lottery, a game of cards, a river).
5. A college freshman is like a rat in a maze.
6. Learning a foreign language is like learning to play golf.
7. Saturday night in the dorm is like winter on the prairie.
8. Starting college is like learning to swim.

Supplementary topics for practice in paragraph development:

1. Develop a short three-paragraph theme on one of the following topics. Try to use a different method of thought development for each paragraph.

   a. On Working Your Way Through College
   b. On Learning to Be a Soldier
   c. On Learning to Be a Coed

    d. Life As a Clerk (or Gas Station Attendant)

    e. What Is a Scholar?

2. Develop a short theme on one of the following "problem and solution" topics in three paragraphs. In the first paragraph, define the problem by means of statistics or details; in the second paragraph, show its causes and effects; in the last paragraph, classify the approaches that might be taken to solving it.

    a. The Drop-out Problem

    b. The Problem of Increasing College Enrollments

    c. My Need for a Larger Budget

    d. Finding a Place to Live While in College

    e. Getting Married While in College

3. Develop a descriptive paragraph on one of the following topics using both time arrangement and space arrangement concurrently.

    a. Approaching the Quad

    b. From Town to the Campus

    c. The A & P

    d. How to Find the Book You Want

3

Writing Essay Examinations

In college you will have to take many essay examinations. An essay examination is one in which you answer questions in your own words. It tests not only your knowledge and understanding of the subject but also your skill in reading and writing. It tests your skill in reading in that it requires you to find out *exactly* what the questions *mean*, not approximately with what they are concerned. It tests your skill in writing in that, if you are to be successful, you must make your meaning unequivocally clear, you must employ a reasonable organization with sufficient thought development, and you must make every word count. And all this you must accomplish in what is essentially a first draft. No wonder, then, that it is difficult to write a good essay examination. There are, however, a few recommended methods and procedures that you can follow to advantage in taking essay examinations, and these are presented below to show you how to cope with your college tests.

BEFORE THE EXAMINATION

Before you attempt to write an essay examination you must have a reasonable command of the subject. Of course, you cannot be expected to know every detail, but you should have in mind an orderly view of the subject, and you should know well those parts of the subject that have received emphasis in class. These two requirements are minimum; from here on you should fill in as much as your time permits. Many students have found it helpful to make out the examination that they would give if they were the instructor, and then to write out the answers to their questions. Sometimes, too, students can work usefully in pairs, asking and answering questions. They usually begin with broad overall questions and then work down to finer details. The important thing in this kind of review session is never to lose sight of the total organization, the view of the material, for this organization helps you to remember and to fit details into their proper place.

DURING THE EXAMINATION

When the examination hour is at hand and you receive the questions, there are three things to bear in mind.

1. Read through the entire examination, both directions and questions, before you begin to write. Note whether a certain

number of minutes is allotted to each question. If so, you have a rough guide as to how much you are expected to go into detail. But if such an allotment of time is not made, you will have to decide for yourself about the relative importance of each question. In either case, plan to complete the examination. Note also whether you are given a choice of questions to answer. If you are given a choice, take it. Do not go ahead and answer all the questions anyway, for only one answer will count, and you will merely waste valuable time with your extra answers.

2. Answer the easiest questions first. This will give you a mental warm-up, and the things you know well may suggest others that you might have overlooked if you had started cold on the hard questions. When you begin with other than the first question, leave plenty of space on your paper for the questions you will answer later, and keep the order of your answers the same as that of the questions.

3. Plan to spend a few minutes at the end rereading your entire paper to tidy up the phrasing, check spelling and punctuation, and add any details that you may have inadvertently omitted.

WRITING THE EXAMINATION

When you are ready to begin writing you will find certain procedures helpful.

1. Note carefully the directive **verb** that tells you what you should **do** in your answer. Directive verbs that are commonly used in examinations are: *explain, enumerate, list, name, compare, contrast, describe, summarize, outline, apply, justify, defend, account for, sketch, clarify, state, illustrate, discuss.* When your instructor uses one of these verbs you may be sure that he means exactly what he says. If, for example, he asks you to *enumerate* or *list* ten migratory birds that pass through Iowa, he expects you only to make a list of their names. He does not expect you to chart their migrations or to describe their appearance or to classify them. If you do give this added information, it will add nothing to your examination grade and will deprive you of time that you need for other questions.

2. Outline and preplan your answer if it is to be of any

length at all. For this purpose use a piece of scratch paper. This preplanning will help you to write an organized instead of a haphazard answer.

3. Stick to the question. Marshal the information you have that is directly relevant to the question and present it in an orderly way. Resist the temptation to make a sly transition to something you know well and to go all out on this area of knowledge. You will not fool anybody, least of all your instructor.

4. Begin your answer with a general statement or topic sentence and develop this according to the methods of thought development you learned in the preceding section entitled "Developing Thought in Paragraphs." This technique works well with many discussion questions, though not with all questions.

You should also be aware of the following caveats in answering examination questions:

1. Do not repeat in other words what you have already said. Suppose, for instance, that you are confronted with this question on a humanities examination: "Contrast the philosophy of the Golden Age in Greece with that of the Hellenistic period." Your answer might well begin with a general statement or topic sentence like this: "Hellenistic philosophers had generally less faith in the power of reason than the Athenian philosophers during the Golden Age." This is a good beginning which you can develop. But here is the way a thoughtless student might continue the answer: "The Hellenistic philosophers were less rational than the Golden Age philosophers. They did not think that reason could solve all of man's problems, while the philosophers of the Golden Age believed that man was capable of arriving at answers to basic problems by using his power of reason." In this answer nothing has really been added to the opening general statement; the second and third sentences merely repeat in other words what has already been said.

2. Do not digress into material that does not answer the question. As an example let us look again at the same question: "*Contrast* the philosophy of the Golden Age in Greece with that of the Hellenistic period." A student might begin with an ac-

ceptable general statement: "The philosophers of the Golden Age in Greece had greater faith in the power of reason than those of the Hellenistic period." But suppose that he continued by naming all the philosophers of the two periods, giving their dates and the schools to which they belonged, but not telling what they believed. Or suppose he wrote a paragraph telling why the Hellenistic philosophers had less faith in reason, speaking of their disillusionment following the Peloponnesian Wars, the ever-present fear of Persia, the conquests of Macedonia, the economic hardships of the masses, and so on. Both of these answers would be digressions: they do not answer the question, "*Contrast the philosophy of . . . .*"

3. Do not use language that is too broad and general. Here is a question that calls for specific details: "*Describe* briefly the three orders of Greek architecture." An answer in this vein would be worthless: "In Greece there were three orders of architecture. Each one had its own characteristics, its own particular style, and its own distinctive claim to beauty. The three varied considerably in many details and were modified in some degree according to the type of building in which they were employed."

4. Do not bluff. The attempt to use elegant but empty language to conceal ignorance never works, and it can be detrimental to you in two ways. First, it wastes time that you might profitably use on other questions. Second, it may irritate your instructor—and it should—because it suggests a lack of intellectual honesty.

PRACTICE EXAMINATION EXERCISE 1

This is the first of two examination exercises designed to help you learn to write better essay examinations.

STEP A.   Study carefully the essay below, "On Dealing with Stereotypes," and write out the answers to the following questions, consulting the essay to make sure that your answers are as nearly perfect as you can make them.

1. (*20 points*)   Define and illustrate stereotypes. (Note

the two directive verbs and be sure that your answer contains both a definition and at least one illustration.)

2. (*30 points*)   Discuss the harm that stereotypes can do. (Do not be deceived by the directive verb *discuss*, which seems to open wide the door to anything you wish to pour out on the subject. Actually, the essay mentions four ways in which stereotypes can do harm, and your instructor will expect you to include these four in your answer, whatever else you might wish to say.)

3. (*40 points*)   Explain how we acquire stereotypes and show how some become deeply ingrained and get a real grip on us. (This is the fattest question, with 40 points, so prepare to do yourself justice here. Note that there are two parts. The answer to this question lies in different parts of the essay, and you will be expected to put these parts together in a single coherent explanation.)

4. (*10 points*)   Describe, in one sentence, the two identifiable methods by which schools deal with stereotypes. (Be sure to follow the one-sentence limitation here. If you do not, you may be downgraded for slovenly reading.)

Now as you go ahead and study the essay you will observe that its organization is not as tight and lucid as you might wish. Make the best of it, keeping in mind that plenty of your college reading will be equally intractable. But the meat is there, and in your class discussion of the answers there should be no doubt about what the answers should contain for full credit.

ON DEALING WITH STEROTYPES

Reading and discussion of books will inevitably reveal that students bring to their experiences in literature some fixed and rigid ideas about groups of people and their characteristics. Teachers need to develop programs which will help children not only to deal with these stereotypes, but also to expand their ideas about people and become more aware of the blocks which stereotyped thinking produces.

Walter Lippmann pointed out, some twenty years ago, how badly distorted our thinking often is by the false images we have of other people. Ever since, many of us have felt eager to eradicate such distortions and stereotypes from our thinking, or at least to immobilize them. . . .

Teachers need to know about stereotypes as one element in conditioning the attitudes of their students; they need to investigate the particular stereotypes which their students learn in their family and community life, and they need to plan a school program which will help their students to deal with these stereotypes as false or inadequate generalizations. Actually, stereotypes are derived from limited experience. Instead of first hand knowledge, the general culture in which we have grown up has provided us with capsule interpretations of people's motives and behavior which most of us have been naïve enough to swallow whole.

As a matter of fact, most of us do not know how ignorant we are of people who live in the same community that belong to different economic, religious, social, or racial groups from ours. We have gathered impressions of them since early childhood from seeing them around or from very limited contacts, and these impressions have often become set mental images—"pictures in your head," as Walter Lippmann would say. Only too easily these snap judgments have been suggested by our associates, friends, and family members in the first place and then reinforced by popular fiction, advertisements, cartoons, stage jokes, hearsay, and the like. Thus, we sometimes have quite clear images of the "lazy Mexicans," "shanty Irish," "dumb Negroes," "scheming Jews," or "sly Japs," although we may have had few, if any, occasions to know individuals from these groups as fellow human beings.

It is these set impressions or pictures-in-your-head that are usually called "stereotypes." In one way or another, they enter into very nearly all our thinking and are applied to many persons with whom we are not in daily, intimate contact. Most of us are aware of such stereotypes because we have found ourselves at one time or another circumscribed or categorized by them. What teacher has not been greeted by child or parent with the remark, "But you don't act like a teacher!" What teacher has not at one time or another been forced by community pressure to dress or entertain or talk like a schoolmarm? We probably also know the popular assumptions about social workers, policemen, lawyers, and workers—especially if they belong to a union. We are familiar with the notions that the rich are usually idle or dissolute, and that the younger generation is still bent on going to the dogs as it is believed to have been doing ever since the world began.

There might not be any very serious harm in snap judgments and false pictures of this kind were it not that they can come between us and reality. For example, consider the country "hick" who has so often amused us in the movies and the funnies. When rural people have come to live in urban centers, they have sometimes found their neighbors reacting to them in terms of their reputation as "hillbillies" or "green-horns." The story is told of certain New York students, who volunteered to help harvest the crops during the wartime shortage of manpower, that for weeks they could not adjust to the necessity of learning from farmers how to do things. They could not make connections with rural life because of their assumptions that all country folk were uncouth, dull, unskilled, unsophisticated, and behind the times. Some of the country people may have been led astray quite as much by their fear that all city people are "slick."

It is important to emphasize that these false pictures and distortions often contain some element of truth. Stereotypes would not be half as misleading as they are if they did not carry a measure of conviction because of something definitely recognizable about them. They are to the mind precisely what cartoon caricatures are to the eye. Stereotypes may accordingly be said to derive from the all but universal habit of quick generalization from inadequate information. We quite naturally prefer slogans and "ten easy lessons" to reasoned analysis, and we try to write so that "he who runs may read." From the educational viewpoint the dependence upon stereotypes and other pattern thinking amounts to virtual abandonment of man's best gift for shaping his life—the power really to use his mind. Furthermore, dependence upon stereotypes prevents the development of the habit of constantly re-examining our generalizations about people. Even if the consequences for human relations were not so apparent and harmful, we would still need to develop ability to interpret new situations in terms of their potentialities instead of accepting stereotypes. Education is obligated to address itself to this task of constantly revising our interpretations and analyses.

But stereotypes can have a more damaging effect. Sometimes the stereotype grew out of a trait or situation that prevailed in the past but which is now undergoing change. The distortion may reflect what one group wants to believe about the other in order to justify its own traditional place in society. Thus, Uncle Tom or the mammy in *Gone with the Wind* may be fairly accurate historical portraits. They become hated symbols only when the dominant culture insists that they represent "the Negro's proper place," or when they are cited as evidence of the "benevolence" of slavery in the Old South. Stereotypes of this category not only reflect limited experience, but also they are used in defense of discrimination. Such stereotypes are the false images about which we are most likely to be unreasonable and defensive, especially when the suspicion may have arisen in our own minds that perhaps they are not as fair and true as we wish they were.

Pictures-in-the-head which have these emotional connotations usually get their real grip on us during childhood training. From infancy we are taught our "appropriate" social roles, as well as the manners, conduct, and dress that are suitable for people like us. Parents frequently compare or contrast the behavior they advocate with that typically associated with other people in the population whose standing is known to be undesirable. The little girl who wants a cerise sweater may be told, "But that is Indian pink; do you want to look like the old squaw we saw in the mountains last summer?" Other children may be admonished not to "act harum-scarum in Sunday school like little wild Indians," or to take a bath so that "teacher will not think you come from a Dago family!" Such parents are teaching their children more than good taste in dress, good manners in church, or cleanliness. The tone of voice, the particular words chosen, clearly indicate that the children from the groups used as bad examples are not wanted around. Stereotypes acquired in this fashion are often so deeply ingrained that most of us are not conscious of how or when we first got them. Any question or challenge of them is, accordingly, like digging into basic assumptions and the very foundations of life.

Lillian Smith has given us a classic description of the way learning about people in subservient roles takes place in the family.

Adapted with permission from *Literature for Human Understanding* (Washington, D.C.: American Council on Education, 1948), pp. 13–18.

It is not easy to pick up such a life and pull out of it those strands that have to do with color, with Negro-white relationships, for they are knit of the same fibers that have gone into the making of the whole fabric; they are woven into its most basic patterns and designs. The mother who taught me what I know of tenderness and love and compassion taught me also the bleak rituals of keeping the Negro in his place. The father who rebuked me for an air of superiority toward schoolmates from the mill settlement, and rounded out his rebuke by gravely reminding me that "all men are brothers," also taught me the steel-like inhuman decorums I must demand of every colored male.

Neither the Negro nor sex was often discussed in my home. We were given little formal instruction in these difficult matters, but we learned our lessons well. We learned the intricate system of taboos, of renunciations and compensations of manners, voice modulations, words, along with our prayers, our toilet habits, and our games. I do not remember how or when, but I know that by the time I had learned that God is love, that Jesus is His Son, that all men are brothers with a common Father, I also knew that I was better than a Negro, that all black folk have their place and must be kept in it, and that a terrifying disaster would befall the South if ever I treated a Negro as my social equal.

"Growing into Freedom," *Common Ground*, IV (Autumn 1943), 47–52.

While teachers do not need to know the whole technical analysis concerning stereotypes, their recognition of these points will be helpful in planning a class or school program: that stereotypes exist in our culture; that they enter into

nearly all of our thinking; that they are rigid, emotionally reinforced generalizations not easily revised in the light of new
experience and information; that they come between us and
reality; that they contain some element of truth; that they
may represent holdovers from the past or are rationalizations
of the wishes of people; that they are learned and underlined
in the family and other group situations where status, affection,
and understanding of role are important determinants of their
feeling content. School programs will more successfully under-
mine those stereotypes which are produced by limited experience and inadequate ways of thinking—in short, those that are
verbalizations without support from social pressures. Such
programs will need to attack more vigorously and with more
thought-out strategy those stereotypes which are reinforced
by patterns of community status, practices of discrimination,
and social prejudice. The degree to which these stereotypes
yield to treatment in school depends on a good many factors
outside the teachers' control.

There are two identifiable methods of dealing with stereotypes. Some schools have taken from their shelves, or refused
to purchase, books containing those false images which are a
source of particular objection and grievance to some minority
groups. Other schools have built units of study and reading
around the positive contributions to American life of different
ethnic strains. It will be worth while to examine both of these
methods in some detail.

STEP B.   You will find below a series of answers that college
freshmen have given to these same questions. Study these
answers and give each one the number of points that you think
it deserves. In each case where you give less than full credit,
be able to justify your grading; that is, be able to point out just
where and how the writer of the answer went wrong, consider-
ing especially what you learned in "Writing Essay Examinations."

1. (*20* points)   Define and illustrate stereotypes.

ANSWERS

    1. Stereotypes are rigid, emotionally reinforced generali-
zations about groups of people and are derived from

lack of experience. An example of a stereotype is the notion that all Mexicans are lazy. Another example is the generalization that all Jews are overly money conscious.

2. Stereotypes are fixed and rigid ideas concerning certain groups of people and their characteristics. An illustration of this is the fixed idea that whites are much better and in a higher class than blacks. They consider themselves better in every respect.

3. A stereotype, by the author of the pamphlet, is one who has a set picture in his mind about certain things in general or one thing in particular. A stereotype usually knows little about the subject but has heard an opinion from an acquaintance or friend. The lack of knowledge is usually the main cause for a stereotype to develop.

4. Stereotypes are very common and all people are likely to have them. Some common stereotypes are "lazy Mexicans," "shanty Irish," "scheming Japs," and "dumb Negroes." It is probably safe to say that there is no one who does not possess a good many of such stereotypes.

5. Stereotypes are "mind pictures." They are ideas that people get about another group of people or things. The majority of these stereotypes are not true. There is just enough suggestion of truth that people tend to believe the whole thing rather than separating the "fact from fiction." Here are some illustrations of stereotypes: (1) The Jew who is "stingy," getting every cent's worth and maybe cheating a little bit. (2) The dumb, dirty Negro. (3) The old maid schoolteacher who is expected to dress in black, wear her hair in a bun, spend her nights reading and associating only with other old maids. (4) The kids from the poorer section of town who are always classed as "juvenile delinquents." (5) The rich people who are always complete snobs and never do anything worthwhile. (6) The "absent-minded professor."

I could go on enumerating the common stereotypes but

they would all prove the same point. That point is that very few people are actually like the stereotype they are supposed to be and even the ones that are a little bit are not enough in number to support such a stereotype.

2. (*30 points*)   Discuss the harm that stereotypes can do.

ANSWERS

1. Stereotypes can cause the loss of the greatest power of our minds, that of thinking. If a person has let stereotypes take over in his mind, it will function thus: "This" is true because I learned it as a child. "That" is true because Jim Jones said so. Therefore I need not think since I know everything anyway.

    Stereotypes can cause us to stop changing our generalities. Once we have let stereotypes take over we no longer change our minds.

    Stereotypes can come between us and reality. For instance, we are brought up on the farm with the idea that all city people are "slickers" and none of them can be trusted. If we then are in the city, we will think that all city folks are bad even if some have more goodness in them than do some rural folks.

2. Stereotypes are harmful because they come between man and reality. They rob him of his power to think and reason. Also, stereotypes may serve to defend and protect practices which are not entirely good for, or beneficial to, mankind as a whole.

3. Stereotypes can be very harmful in that they often give false ideas and conceptions of certain things. An example of this would be to say: "White people are superior to dark people." This is not true and leads you to believe that those with dark skins are less intelligent than those with lighter skins. An example of how this misconception could be harmful might be seen in this way.

Suppose you were an employer and you were interviewing applicants for a certain job. The applicants have been eliminated until there are only two left, one with light skin and one with dark. The person with the dark skin may have more training in the specific job and therefore be more suited for the position than the lighter skinned person. But the job is one of prestige and because of the belief in the stereotype that the whites are more intelligent than darks, you place the light skinned person in the position instead of the dark, even though the dark skinned person is better qualified.

This is just one example of how a stereotype can be harmful. Stereotypes lead us to believe things about people that are not true. By believing these things we judge accordingly and our judgment of a person can often be false because of the stereotype.

4. Stereotypes harm religious, political, racial, and social groups both locally and nationally. They can cause bad feelings between friends, towns, or even nations. People have been known to kill because they had been taught that the victim was useful to no one and should therefore be disposed of. A feeling of belief toward a person or thing can harm an individual emotionally. For instance, a person may be pushed to the point of a nervous breakdown because, contrary to his childhood teachings, he likes a group of people, for example, Negroes or Jews. It may prey on his mind that he is abnormal or something. Stereotypes harm us mentally because they stop the healthy, normal process of thinking for ourselves. We hear something and instead of investigating we simply accept it and continue on.

5. Because stereotypes are not wholly truthful, some people get false impressions about other people or things. This is very harmful because a person might have nothing to do with someone whom he thinks he is superior to. A false impression like this could even go so far as to build up hate in the man. It could also make the person lose many friends and acquaintances.

3. (*40 points*)   Explain how we acquire stereotypes and show how some become deeply ingrained and get a real grip on us.

ANSWERS

1. Most of us acquire our use of stereotypes quite young from our parents, relatives, and so on. Then, as we grow older, we become influenced by such things as motion pictures, advertisements, and books. Our friends also have a big part in influencing us. After continual use of these false images they become an actual part of our thinking.

2. We acquire most of our stereotypes during childhood. Our folks may often try to get us to become better socially by telling us not to be like the dirty Mexicans, or not to dress gaudy like the blacks. All of these are generalizations that are acquired from lack of knowledge, experience, and understanding. When we go shopping, our mother might tell us not to go *there* because they are Jews. In some communities that are mostly Protestant, Catholics might be called "fish eaters," and this forms a stereotype in the children's minds.

   The stereotypes that really get a grip on us are the ones that are evident in all institutions in our particular community. The ones that are just generalizations taken out of a book are not as hard to correct and control. In my community everyone formed stereotypes of Bahamans and Mexicans. The reason we were taught that these people were bad, lazy, and dirty was because every summer some would be shipped into our town to do detasseling. Naturally these were the worst ones. We were told not to talk to them and to stay in the house. Anyone who would talk to these detasselers would be gossiped about. This led many children in our community to think of *all* Mexicans and Bahamans as *bad*. This is deeply imprinted in our lives because we can think back and remember the incidences of the detasselers. The inferiority of the Mexicans and Bahamans was very evident in our school, family, and other organizations.

3. Everyone acquires stereotypes as he goes through life. Usually they are about people with whom he has had little experience. For instance, the man who has had little education may have the stereotype of the old-maid schoolteacher with a black dress, high shoes, and a cross disposition. The person who has never been in the military service will probably have a stereotype of a big, burly, tough army master sergeant who speaks foul language. The city boy who has always lived in the city may have the "redneck" or "hick" stereotype. When such stereotypes become deeply ingrained, they are dangerous to us because they come between us and reality, they keep us from seeing people as they really are.

4. Stereotypes are acquired most often in the pattern-forming childhood years. At an early age children accept readily what their parents and others may tell them or imply by certain remarks.

Thus, when a mother tells her children to wash well or the schoolteacher will think you come from a Dago family, the children get the idea that Dago children are not desirable. Inferences such as these are repeated and stressed more and more as the children get older and become a part of the way they think. Social pressures are brought to bear on people who do not accept a stereotype that the community believes. For example, a white man who hires a black for a top office job might face expulsion from the groups he has associated with all his life and possibly threats to the well-being of his family or person. After seeing these pressures brought against others by all his friends, a person can only feel that the stereotype must be true. No matter how much new experience or information is acquired concerning the streotype, the person will still believe what has become ingrained in his mind all through his life.

5. Stereotypes are acquired in many ways but most often they are implanted on our minds by our parents, relatives, or close friends. Sometimes we acquire stereotypes through association with someone of this sort, and we have the same feeling toward all those who are like him.

Stereotypes become deeply ingrained and get a real grip on us because we see them almost every day in advertisements, comic strips, and magazines, or we hear them from various sources such as parents, radio, or television. After we hear and see these stereotypes so often, we feel that all people of that nationality, race, religion, or vicinity must be the same. Again, whites' attitude toward blacks is a good example. Some whites have been raised on the idea that all blacks are ignorant and militant.

4. (*10 points*)   Describe in one sentence the two identifiable methods by which schools deal with stereotypes.

ANSWERS

1. It is very necessary that schools deal with stereotypes by identifiable methods because our country has many minority groups and these will suffer if people think of them in stereotypes. For example, if we hold in our minds the stereotypes of the "crafty Jew," the "vicious Negro," the "dirty Indian," then when we meet an honest Jew or a gentle Negro or a clean Indian we shall misjudge them. And the school is the best place to eradicate such undesirable stereotypes because the school gets children young and can implant the right ideas early so they have a chance to grow.

2. Schools may try to combat stereotypes by keeping from the children such books or pictures that might help to develop stereotyped images, and by giving children a full and true picture of the customs, background, and contributions of groups who are usually victims of stereotypes.

3. Stereotypes are dealt with by exposing children as early as possible to books and other examples of the assets, contributions, and admirable qualities of a stereotyped person or group.

   Stereotypes are also dealt with by the removal of all books or other sources of information which tend to increase or intensify the ideas concerning stereotyped people or groups.

4. Some schools do not buy books that may cause prejudice among certain groups; others build a field of study around unethical beliefs.

5. Two methods by which schools deal with stereotypes are: first, usage and discussion of material concerning stereotypes; second, avoiding stereotypes as much as possible.

PRACTICE EXAMINATION EXERCISE 2

Tomorrow you will take a half-hour practice examination on the selection below. In preparation you are asked to do the following. First, study the material carefully, underlining and marking it to help you locate the central ideas and main facts. Then, write out five precisely worded questions, the answers to which will represent the meat of the selection, and be prepared to answer your own questions. The examination will consist of five questions which your instructor believes will test your knowledge and understanding of the material, and if you take pains to apply your intelligence, you should be able to anticipate most of them. Before the examination you are to hand in your questions. After the examination you will have an opportunity to reexamine and discuss the selection and to analyze your answers.

The selection is taken from a college textbook for freshman courses in the humanities. A single day's assignment will often be from five to ten times as long. This selection, in contrast with the one on stereotypes, is neatly and lucidly organized, as many of your basic textbooks will be.

ATHENIAN LIFE IN THE GOLDEN AGE

The population of Athens in the fifth and fourth centuries was divided into three distinct groups: the citizens, the metics, and the slaves. The citizens, who numbered at the most about 160,000, included only those born of citizen parents, except for the few who were occasionally enfranchised by special law. The metics, who probably did not exceed a total of 100,000, were resident aliens, chiefly non-Athenian Greeks, although some were Phoenicians and Jews. Save for the fact that they had no political privileges and generally were not permitted to own land, the metics had equal opportunities with citizens. They could engage in any occupation they desired and participate in any social or intellectual activities. Contrary to a popular tradition, the slaves in Athens were never a majority of the population. Their maximum number does not seem to have exceeded 140,000. On the whole, they were very well treated and were often rewarded for faithful service by being set free. They could work for wages and own property, and some of them held responsible positions as minor public officials and as managers of banks.

Life in Athens stands out in rather sharp contrast to that in most other civilizations. One of its leading features was the amazing degree of social and economic equality which prevailed among all the inhabitants. Although there were many who were poor, there were few who were very rich. The average wage was the same for practically all classes of workers, skilled and unskilled alike. Nearly everyone, whether citizen, metic, or slave, ate the same kind of food, wore the same kind of clothing, and participated in the same kind of amusement. This substantial equality was enforced in part by the system of liturgies, which were services to the state rendered by wealthy men, chiefly in the form of contributions to support the drama, equip the navy, or provide for the poor.

A second outstanding characteristic of Athenian life was its poverty in comforts and luxuries. Part of this was due to the low income of the mass of the people. Teachers, sculptors, masons, carpenters, and common laborers all received the same standard wage of one drachma (about 30 cents) per day. Part of it may have been due also to the mild climate, which made possible a life of simplicity. But whatever the cause, the fact remains that, in comparison with modern standards, the Athenians endured an exceedingly impoverished existence.

They knew nothing of such common things as watches, soap, newspapers, cotton cloth, sugar, tea, or coffee. Their beds had no springs, their houses had no drains, and their food consisted chiefly of barley cakes, onions, and fish, washed down with diluted wine. From the standpoint of clothing they were no better off. A rectangular piece of cloth wrapped around the body and fastened with pins at the shoulders and with a rope around the waist served as the main garment. A larger piece was draped around the body as an extra garment for outdoor wear. No one wore either stockings or socks, and few had any footgear except sandals.

But lack of comforts and luxuries was a matter of little consequence to the Athenian citizen. He was totally unable to regard these as the most important things in life. His aim was to live as interestingly and contentedly as possible without spending all his days in grinding toil for the sake of a little more comfort for his family. Nor was he interested in piling up riches as a source of power or prestige. What each citizen really wanted was a small farm or business which would provide him with a reasonable income and at the same time allow him an abundance of leisure for politics, for gossip in the market place, and for intellectual or artistic activities if he had the talent to enjoy them.

It is frequently supposed that the Athenian was too lazy or too snobbish to work hard for luxury and security. But such was not quite the case. It is true that there were some occupations in which he would not engage, because he considered them degrading or destructive of moral freedom. He would not break his back digging silver or copper out of a mine; such work was fit only for slaves of the lowest intellectual level. On the other hand, there is plenty of evidence to show that the great majority of Athenian citizens did not look with disdain upon manual labor. Most of them worked on their farms or in their shops as independent craftsmen. Hundreds of others earned their living as hired laborers employed either by the state or by their fellow Athenians. Cases are on record of citizens, metics, and slaves working side by side, all for the same wage, in the construction of public buildings; and in at least one instance the foreman of the crew was a slave.

In spite of expansion of trade and increase in population, the economic organization of Athenian society remained comparatively simple. Agriculture and commerce were by far the most important enterprises. Even in Pericles' day the majority of the citizens still lived in the country. Industry was not highly developed. Very few examples of large-scale production are on record, and those chiefly in the manufacture of pottery and implements of war. The largest establishment that ever existed was apparently a shield factory owned by a metic and employing 120 slaves. There was no other more than half as large. The enterprises which absorbed the most labor were the mines, but they were owned by the state and were leased in sections to petty contractors to be worked by slaves. The bulk of industry was carried on in small shops owned by individual craftsmen who produced their wares directly to the order of the consumer.

Religion underwent some notable changes in the Golden Age. The primitive polytheism and anthropomorphism of the Homeric myths were largely supplanted, among intellectuals at least, by a belief in one God as the creator and sustainer of the moral law. Such a doctrine was taught by many of the philosophers, by the poet Pindar, and by the dramatists Aeschylus and Sophocles. Other significant consequences flowed from the mystery cults. These new forms of religion first became popular in the sixth century because of the craving for an emotional faith to make up for the disappointments of life. The more important of them was the Orphic cult, which revolved around the myth of the death and resurrection of Dionysus. The other, the Eleusinian cult, had as its central theme the abduction of Persephone by Pluto, god of the nether world, and her ultimate redemption by Demeter, the great Earth Mother. Both of these cults had as their original purpose the promotion of the life-giving powers of nature, but in time they came to be fraught with a much deeper significance. They expressed to their followers the ideas of vicarious atonement, salvation in an after-life, and ecstatic union with the divine. Although entirely inconsistent with the spirit of the ancient religion, they made a powerful appeal to certain classes of Greeks and were very largely responsible for the spread of the belief in personal immortality. The majority of the people, however, seem to have persisted in their adherence to the worldly, optimistic, and mechanical faith of their ancestors and to have shown little concern about a conviction of sin or a desire for salvation in a life to come.

It remains to consider briefly the position of the family in Athens in the fifth and fourth centuries. Though marriage was still an important institution for the procreation of children who would become citizens of the state, there is reason to believe that family life had declined. Men of the more prosperous classes, at least, now spent the greater part of their time away from their families. Wives were relegated to an inferior position and required to remain secluded in their homes. Their place as social and intellectual companions for their husbands was taken by alien women, the famous hetaerae, many of whom were highly cultured natives of the Ionian cities. Marriage itself assumed the character of a political and economic arrangement, devoid of romantic elements. Men married wives so as to insure that at least some of their children would be legitimate and in order to obtain property in the form of a dowry. It was important also, of course, to have someone to care for the household. But husbands did not consider their wives as their equals and did not appear in public with them or encourage their participation in any form of social or intellectual activity.

4

Organizing by Outline

The scratch outline described in the beginning pages is usually all you need in preparing to write a short informal theme. But as your theme assignments grow longer and more complex, you may find it handy to learn how to make a more careful and detailed outline. Though at first this may seem merely an added chore, it will really save you time and should result in better themes and grades.

An outline is the ground plan for a piece of writing. It consists of a series of sentences or topics in an orderly sequence and grouping. Practice in outlining is useful to a writer because it increases skill in organizing ideas.

Your outline helps you to arrange your main points in order, to group supporting material under the proper main points, and to place the parts you wish to emphasize in the most strategic positions.

Your outlines for papers will often be short and simple. If, for example, you prepare an outline of a short theme, it might be a simple one like this:

CENTRAL IDEA: Underlining is a valuable study aid to a college student.

1. *Underlining forces you to think.*
   a. It helps you to find main points.
   b. It shows the thought structure of the assigned reading.
   c. It helps you to associate examples and supporting material with the right points.

2. *Underlining enables you to review quickly before an examination.*
   a. You can quickly locate the most important general statements and ideas.
   b. You can see at once what supporting material you need to restudy.

This outline contains your two main points supporting the central idea and the supporting subpoints under each main point—all in the order in which you intend to present them. Notice that each item is a full sentence. Using full sentences in an outline helps you in one important respect: it shows you that you have definite *points* to discuss, not just vague topics. But, if you choose, you may use topics instead of sentences. Here is a topic outline that you might use for a short paper:

CENTRAL IDEA: Knowing how to review will give you better examination grades.

1. *How to review alone*
   a. Finding the organization of each block of reading
   b. Finding the major points (ideas or general statements)
   c. Filling in with supporting points or details
   d. Correlating class notes with reading

2. *How to review with a classmate*
   a. Deciding what is most important
   b. Finding relationship of most important points
   c. Questioning each other and verifying answers
   d. Comparing class notes

An outline is a private affair. No one sees it but the writer because it is just a means to an end, the end of whipping your material into a logical and effective shape. Therefore, you do not have to be fussy about most matters of form—punctuation, capital letters, the system of numbering and lettering, and so on. One matter of form, however, does require careful attention: all items in the same sequence of numbers or letters should be in parallel form, that is, they should have the same grammatical structure. Parallel form guards your organization. It helps you to make sure that the ideas you have grouped together really belong together, and it keeps you from straying off on an irrelevant point. As an example, let us look again at point two of the outline above, with one item thrown out of parallel form:

2. *How to review with a classmate*
   a. Deciding what is most important
   b. Finding relationship of most important points
   c. Questioning each other and verifying answers
   d. Class notes

In this numbered sequence, *class notes* is not parallel; it is a noun, and the three other items begin with verbals indicating an action. Now, with the term *class notes* in your outline, the question for you as a writer is, "What about class notes?" What are you going to say about them? Will you tell how il-

legible they are? Will you describe the evident omissions? Will
you advise your readers to query the lecturer about doubtful
points? If you do any of these things, you will be straying off
into an irrelevancy, and your readers will notice this at once.
But if your outline form is parallel—*d. Comparing class notes*
—you will know that you are to write about an action that you
and your classmate will engage in, an action that is in line with
the three previous actions—*deciding, finding,* and *questioning.*
And your readers will perceive that you are adding a fourth
point that is parallel in idea with the preceding three.

If your outline is really going to assist you in planning a
paper, it must be specific, informative, and full. Such an outline will channel your thoughts smoothly from beginning to
end, and you can then concentrate on filling in details and
making effective transitions. The usefulness of an outline that
is specific, informative, and full can be easily demonstrated.
Below, you will find two outlines for a paper on this central
idea: "I plan to be an elementary teacher." The first one, you
will see, is vague and barren, and would be of little assistance
to you in preparing your paper. When you read the second one,
compare it point by point with the first.

FIRST OUTLINE

1. *Introduction*
   a. What my friends say
   b. How I react toward children

2. *Why go into teaching?*
   a. Service to community
   b. Work
   c. World
   d. Vacations

3. *Why avoid teaching?*
   a. Work
   b. Money

4. *Conclusion*

SECOND OUTLINE

I. *My interest in teaching*
   A. My friends say it is personally satisfying
   B. I like children

II. *Rewards of teaching*
   A. Opportunity for useful service
      1. Character is formed in early grades
      2. Habits leading to later success are acquired in first two grades
   B. Regular employment
   C. Chance to see world
      1. Primary teachers in demand in many large cities
      2. Government opportunities to teach abroad
         a. In military services
         b. On Fulbright program
         c. In noncontinental U.S.
      3. Chance to hook a man
         a. Will meet many men while traveling
   D. Long vacations

III. *Disadvantages of teaching*
   A. Tiring work and long hours
   B. Low pay and retirement benefits

IV. *Rewards exceed disadvantages*

The second outline is beginning to look usable; it is more specific, informative, and full. But before you begin to write from it, you had better look at it with a critical mind and ask a few key questions.

1. Do I have a reasonable and effective order to my points? As we look at the outline, we notice that the major point seems to be "Rewards of teaching" and that this is in the center, which is not a position of emphasis. Wouldn't it be better to follow the introductory first point with "Disadvantages of teaching"? You could lead into it with a transitional sentence like, "It is true, of course, that a teaching career does have some disadvantages." Then, after discussing these disadvantages, you could move to your major point as the culminating

section, using a transitional sentence such as, "Though teaching has its disadvantages, it also offers rich rewards."

2. Do I have enough to say about each point? As we study the outline, two points seem rather thin: "Regular employment" and "Long vacations." Neither has any subpoints, and there really isn't much to say about either one. Perhaps they could be combined into one: "Regular employment with long vacations." Now the point is a little more substantial.

3. When I use subpoints, are there always at least two? The reason for the requirement of at least two subpoints is a logical one: you cannot divide anything into fewer than two parts. If, then, you have only one subpoint, it is probable that it should be subsumed under the point above. Looking at the outline, we find one point, "Chance to hook a man," that has only one subpoint, "Will meet many men while traveling." So let us combine them in this way: "Chance to meet men and hook one."

This is how the outline will look:

CENTRAL IDEA: I plan to be an elementary-school teacher.

I. *My interest in teaching*
   A. My friends say it is personally satisfying
   B. I like children

II. *Disadvantages of teaching*
   A. Tiring work and long hours
   B. Low pay and retirement benefits

III. *Rewards of teaching*
   A. Opportunity for useful service
      1. Character is formed in early grades
      2. Habits leading to later success are acquired in early grades
   B. Regular employment with long vacations
   C. Chance to see world
      1. Primary teachers in demand in many large cities
      2. Government opportunities to teach abroad
         a. In military services

     b. On Fulbright program
     c. In noncontinental U.S.
  3. Chance to meet men and hook one
IV. *Rewards exceed disadvantages*

Here at last is a usable outline. However, it is still imperfect: it mixes topics with sentences; point IV is not in parallel form; and point C3 does not seem in line with points C1 and C2. Of these three imperfections, the third one may give you trouble, for it is a logical matter. The difficulty is that the *idea* here is not logically parallel with those in C1 and C2, which are both concerned with chances to get jobs; and in writing you will be hard put to make your matrimonial scheme in C3 fit in with the two preceding points on getting a job. There are two solutions to your difficulty. One is to remove point C3, for you still have enough material for your paper. The other is to recast the whole of point C, which could go something like this:

C. Chance to see world

  1. Job opportunities
     a. Primary teachers in demand in large cities in U.S.
     b. Government opportunities to teach abroad
        (1) In military services
        (2) On Fulbright program
        (3) In noncontinental U.S.

  2. Cultural benefits
     a. Learn how other people live
     b. See famous places
     c. Meet men and hook one

Now you have a logical structure, one in which you can easily make a transition from C1 to C2. You can safely ignore the other imperfections in the outline, which are merely formal, and go ahead with the writing. And with such an outline— logical, specific, informative, and full—the job of writing is already half-done.

Let us now start from the beginning and see how an outline is built, step by step. We shall assume that you, a college freshman, have received a letter from the English 12 class of your hometown high school. A number of the class are going to college next year, and they are getting worried about the hazards of college work. So they have asked you to write an article for the school paper on what they should do in English 12 to get ready for freshman English and for college classes in general. This is a golden opportunity for you. Now you must get to work laying your ground plan. And here are the four stages you will probably go through to prepare a usable outline for what you hope will be a dazzling article.

STAGE 1. The first thing to do is to get all your ideas down. Their form and order do not matter yet, for they are only your raw material. Just get them down as they come to your mind. After half an hour of scribbling, you have a list that will look something like this:

1. Better learn to *use* your language.
2. Using language is more important than knowing about literature.
3. College reading is stiff.
4. To read well you should know how to underline and take notes.
5. You've got to remember what's in your reading assignments.
6. Writing standards are high.
7. Better know spelling and punctuation before coming to college.
8. Learn how to write exposition; it's the most important and the hardest.
9. Know good usage for writing.
10. Organizing your papers.
11. Practice with paragraphs.
12. Reading—ask yourself key questions and answer them, and tell someone exactly what you have read.
13. College writing should be accurate—say what you mean.

14. Speaking is learned through practice—get it now.

15. Colleges require public speaking.

16. College writing must be clear at first reading.

17. Must learn to listen to lectures.

18. Must learn to remember what you hear in lectures— important things, that is.

19. Must learn to take notes on lectures.

20. In speaking, organize your talk.

21. Self-confidence through speaking.

22. Learn to deliver a speech smoothly.

23. Speaking will help you to get into debate.

24. Writing will help you get on the college paper.

25. Recreational reading.

26. Much information gotten through college lectures.

27. Freshman composition is required everywhere, of all freshmen.

28. Practice reading hard stuff now.

29. Learn to outline what you read to get general organization.

30. Students fail on account of poor writing.

31. Much writing practice is needed in high school.

32. Learn to write connected sentences.

STAGE 2.   Now you must find a central idea for your article and sort your raw material into groups. At the same time you may want to tidy up the wording of some items, and you can add anything new that you think of. When you have completed this stage, your material will probably look like this:

CENTRAL IDEA: Learn to use your language.

Reading

1. College reading is stiff.

2. To read well you should know how to underline and take notes.

3. You must remember what's in your reading assignments.

4. Ask yourself key questions and answer them.
5. Tell someone exactly what you have read.
6. Recreational reading.
7. Practice reading solid stuff now.
8. Learn to outline for organization.

Writing

1. Writing standards are high.
2. Better know spelling and punctuation before coming to college.
3. Learn to write exposition; very important and hard.
4. Know good usage.
5. Organize your papers.
6. Practice with paragraphs.
7. College writing must be accurate; you must be able to say what you mean.
8. College writing must be clear.
9. Writing will help you get on staff of college paper.
10. Freshman composition is required everywhere of freshmen.
11. Students fail because of poor writing.
12. Much writing practice needed in high school.
13. Learn to write connected sentences.

Speaking

1. Speaking is learned through practice; get it now.
2. Colleges require public speaking.
3. Organize your talk.
4. Develop self-confidence in speaking.
5. Deliver speech smoothly.
6. Speaking will help you get into debate.

Listening

1. You must learn to listen to lectures.
2. You must learn to remember what is important in lectures.
3. You must learn to take good lecture notes.

4. Much information is gained through classroom lectures.

Extras

1. Using language is more important than knowing about literature.

STAGE 3. Now comes the hard work. In each group you must find a reasonable order for your ideas. You must sift out your main points, get them in order, and list the subpoints under them, so far as you can do it at this stage. You can also continue to improve your wording and can add anything further that you chance to remember. When this stage is completed—and it's the hardest one—your article will be taking shape. Here is how it might look at stage three:

CENTRAL IDEA: You should develop your language skills in high school.

1. *Reading*
   a. College reading is difficult.
   b. You must remember what you have read.
   c. You should practice with solid reading now.
   d. Learn to underline and take notes.
   e. Learn to outline for organization.
   f. Learn to ask key questions and answer them.
   g. Retell to others what you read.
   h. Recreational reading???

2. *Writing*
   a. Freshman composition is required in most colleges.
   b. Writing standards in college are high.
      (1) You should know spelling, punctuation, and good usage before coming.
      (2) Many students fail because of poor writing.
      (3) College writing must be accurate.
      (4) College writing must be clear.
   c. You should get much writing practice in high school.
      (1) Exposition is most important.
      (2) Practice writing different kinds of paragraphs.
      (3) Organize each paragraph, with beginning, middle, and end.
      (4) Learn to connect your sentences with transitions.
   d. Writing will help you get on college paper.

3. *Speaking*
  a. Colleges often require public speaking of freshmen.
  b. Get speaking practice now.
    (1) Gives you self-confidence.
    (2) Gives you smooth delivery, free from common faults.
    (3) Gives you practice in organization.
    (4) May help you get on debate team???

4. *Listening*
  a. Much learning is gained from classroom lectures.
  b. You must and can learn to listen effectively to lectures.
  c. Learn to remember what is important—main points and their connections.
  d. Learn to take good lecture notes while listening.

STAGE 4.   Here you do the final arranging and polishing. First of all, you must consider the order of the four groups. Which one is the most important and where do you want it, first or last? If you decide that the "Writing" group is the most important and that "Reading" is the second in importance, you might place "Writing" last and "Reading" first. And if you use these positions, will the other groups fit into place satisfactorily? Then, you have other questions to consider. What is the very best order of main points and subpoints? Should you combine some items and split others? Are there any unneeded points? If so, leave them out. Is the wording of the central idea clear and forceful? Do you have parallel form in each sequence? And finally, what are you going to write for a strong conclusion? When you have settled these questions you might come out with an outline like this:

CENTRAL IDEA: To prepare for college in high school English you should develop your language skills.

1. *Reading*
  a. College reading assignments are long and difficult.
  b. You must *know* in an orderly way what you have to read.
  c. You should practice now with solid reading matter, like that in college textbooks.
    (1) Learn to underline.
    (2) Learn to outline for organization, and to take notes.

    (3) Practice asking yourself key questions and answering them.

    (4) Practice retelling to others the content of what you have read.

2. *Speaking*

  a. Many colleges require public speaking of freshmen.

  b. You should get speaking practice now.

    (1) Gives you self-confidence.

    (2) Helps you to overcome common weaknesses in delivery.

    (3) Teaches you to prepare a well-organized talk.

3. *Listening*

  a. In college much learning comes from listening to classroom lectures.

  b. Your ability to listen effectively can be developed through practice.

    (1) Learn to remember what you hear by noting the main points and their relationships.

    (2) Learn to take good notes while listening.

4. *Writing*

  a. Freshman composition is required in almost all colleges.

  b. College writing standards are high.

    (1) Many fail because of poor writing.

    (2) You are expected to know spelling, punctuation, and good usage.

    (3) College writing must be accurate and clear.

  c. You should practice often, for you can learn to write only by writing.

    (1) Practice hardest on exposition, that is, explaining things.

    (2) Learn to write different kinds of paragraphs, developing your thoughts in various ways.

    (3) Learn to organize your thoughts, with a beginning, a middle, and a conclusion.

    (4) Learn to connect your sentences in thought so that they flow smoothly.

5. *Conclusion*

It is these language skills that pay off in college. Get a head start by developing them now.

Such is the way that useful outlines are made. Remember that the outline is for you alone and that it is serviceable when it enables you to:

1. Arrange your points or topics in the best order.
2. Combine in groups those things that go together.

INDEX

A

Achieving sentence flow, 112–125
Active (use of) instead of passive recommended, 15–17
active-passive shift, 16
Agreement of pronoun and antecedent, 76–77
Agreement of subject and verb, 9–11
Ambiguity, 12–15
Analogy in developing paragraphs, 128, 142
Apostrophe, 68–69

B

Beginning to write, 1–2
Brainstorming, 2, 6–8

C

Capitalization and lower case, 18–21

Cause and effect in developing paragraphs, 128, 140
Central idea in a theme, 1–4, 127, 173
Chronological order in developing paragraphs, 128, 130
Classification and division in developing paragraphs, 128, 138
Cliché, 30–31
Coherence (*see* Achieving sentence flow, Transition, and Sentence beginnings)
Colloquial language, 34–35
Colon, 32–33
and quotation marks, 72
Comma, 21–22, 26–29
and nonrestrictive clauses, 60–62
within quotation marks, 71
and pause, 26, 60–61
after transition words, 86
and vocal behavior, 24, 60–61
Comma fault or comma splice, 23–25

179

Confused sentence, 35–36

D

Dash (use of) recommended, 38–39
   within quotes, 72
Developing thought in paragraphs, 126–144
Diction
   ambiguity, 12–15
   cliché, 30–31
   colloquial, 34–35
   confused sentence, 35–36
   on essay exams, 148
   imprecision, 42–43
   meaning is not clear, 44
   mixed metaphor, 50–51
   nonstandard, 52–53
   redundancy, 75
   repetition, 79–80
   slang, 89–90
   wordiness, 104–105
   wrong word, 106–110
Direct and indirect quotations, 71
Division and classification in developing paragraphs, 128, 138

E

Ellipsis in quotations, 73
Emphasis
   achieving sentence flow, 112–125
   capitalization and lower case, 18–21
   developing thought in paragraphs, 126–144
   parallel structure, 63–64
   periodic structure, 65–67
   repetition, 79–80
   sentence variety, 97–98
   subordination, 95

Essay examinations, 144–164
   preparation for, 145, 162
   procedures during exam, 145–148
   reading questions, 146
Examples (use of) in developing paragraphs, 128–129

F

First draft, 5
Fragment, 40–41
Fused sentence, 81–82

G

General-and-specific in developing paragraphs, 128, 133
Grammar
   agreement, 9–11, 90–91
   person, 90
   reference of pronouns, 76–78
   tense, 101

H

Historical present tense, 101
Hook-and-eye links between sentences, 115

I

Imprecision, 42–43
Indirect quotations, 71
   and tense shift, 101
Inflection of nouns and verbs, 9
Irrelevant material
   in paragraphs, 57–58, 126–127
   in writing exams, 146–148
Italics, 72

K

Key terms, 112–114
   capitalization of, 18

L

Limiting (restrictive) clauses, 60

M

Meaning is not clear, 44
Misplaced modifier, 46–47
Misrelated opening modifier, 48–49
Mixed metaphor, 50–51

N

Nonstandard English, 52–53

O

Organization, 1–8, 165–178
    on essay exams, 147
Organizing by outline, 165–178
Outline
    form, 166–172
    logical structure, 169–170, 175
    scratch outline, 2–4
    sentence outline, 165–166, 169
    for a short theme, 1–8, 166–168
    steps in preparation, 171–178
    topic outline, 1–8, 166

P

Paragraphs, 56–59, 126–143
    irrelevant material in, 57, 126–127
    length of, 126–127
    topic of, 56–57, 126–127
Parallel structure, 63–64
    in outlines, 167
    and sentence flow, 116–118
Passive construction, 15

Patterns of thought development, 127–143
    analogy, 142
    cause and effect, 140
    comparison, 136
    division and classification, 138
    examples, 128–129
    general and specific, 133
    space arrangement, 132
    statistics, 135
    time arrangement, 130
Periodic structure, 65–67
Possessive, 68–69
Practice examinations, 148, 161
Preparation for exams, 144–145, 161
Pronouns, reference of, 76–78
Punctuation
    of adjective clauses, 60–61
    apostrophe, 68
    colon, 32–33
    comma, 21–29
    dash, 38–39
    of a fragment, 40–41
    of interrupters, 27, 38
    and pause, 24, 60–61, 86
    within quotations, 72–73
    of a run-on sentence, 81–82
    semicolon, 86–87
    of a series, 27, 32, 38, 87
    and vocal behavior, 24, 60–61

Q

Quotation marks, 71–74
    for direct quotations, 71–72
    for long and short passages, 73
    with other punctuation marks, 71–72
    for quote within a quote, 71
    for titles, 72

R

Redundancy, 75

Repetition, 79–80
  for emphasis, 79
  on exam questions, 147
  of key terms, 114–115
  of meaning, 75
  of sentence pattern, 97–98
Restrictive (limiting) clauses, 61
Revision of rough draft, 6
Run-on (fused) sentence, 81–82

S

Sample questions and answers
    for essay exams, 148, 161
Scratch outline, 2–4
Semicolon (use of) recommended,
    86–87
  and quotation marks, 72
Sentence beginnings, 85–87
Sentence flow, 112–125
Sentence outline, 165–166, 169
Sentence structure
  active instead of passive (use
    of) recommended, 15–17
  ambiguity, 12–15
  beginnings, 85–87
  confused sentence, 35–36
  imprecision, 42–43
  meaning is not clear, 44
  misplaced modifier, 46–47
  misrelated opening modifier,
    48–49
  overburdened sentence, 53–54
  parallel structure, 63–64
  periodic structure, 65–67
  subordination, 95–96
  transition, 99–100
  variety, 97–98
  wordiness, 104–105
Sentence variety, 97–98
Series, punctuation of, 27, 32, 38,
    87
Shift from active to passive, 16
Shift of person, 90–91
Shift of tense (see Tense shift)
Slang, 89–90

Space arrangement in developing
    paragraphs, 128, 132
Spelling, 92–94
Spelling list, 93
Statistics in developing para-
    graphs, 128, 135
Structural links between sen-
    tences, 85–87, 115–120
Style in capitalization, 18
Subordination, 95–96
Syntax
  active instead of passive (use
    of) recommended, 15–17
  ambiguity, 12–15
  comma fault or comma splice,
    23–25
  fragment, 40–41
  imprecision, 42–43
  meaning is not clear, 44
  misplaced modifier, 46–47
  misrelated opening modifier,
    48–49
  overburdened sentence, 53–54
  periodic structure, 65–67
  punctuation   of   adjective
    clauses, 60–61
  run-on (fused) sentence, 81–
    82
  sentence beginnings, 85–87
  sentence variety, 97–98
  subordination, 95–96
  transition, 99–100
  wordiness, 104–105

T

Tense shift, 101–102
Thought development (see De-
    veloping thought in para-
    graphs)
  on essay exams, 144–148
Time arrangement in developing
    paragraphs, 128, 130
Titles, capitalization of, 18
  in quotation marks, 72
  underlining, 72

Topic sentence, 56–58, 126–127
  in answering essay question,
  147
Transition, 99–100
  between paragraphs, 118
  between sentences, 85–87, 99–
  100, 112–126
Transition words, 99–100
  punctuation of, 86
Transitional devices, 99–100, 112

U

Underlining, 72

Undeveloped paragraphs, 58, 126–
  127

W

Weak passive, 16
Wordiness, 104–105
Words often confused in spelling,
  94
Words often misspelled, 93
Words used as words, 72
Wrong word, 106–110 (*see also*
  Spelling)

CORRECTION

WORD CHOICE:

Ambiguity	2 Amb
Cliché	8 Cl
Colloquial	10 Colloq
Imprecision	14 Imp
Meaning Is Not Clear	15 M
Mixed Metaphor	18 MM
Nonstandard	19 NS
Redundancy	27 Red
Repetition	29 Rep
Slang	33 Sl
Wrong Word	41 WW

SENTENCE CLARITY:

Achieving Sentence Flow	Chap. 3, 1
Ambiguity	2 Amb
Confused Sentence	11 Conf
Imprecision	14 Imp
Meaning Is Not Clear	15 M
Misrelated Opening Modifier	17 Mis O M
Transition	38 T
Wordiness	41 W

SENTENCE FORM:

Use of Active Instead of Passive Recommended	3 AP
Comma Fault or Comma Splice	6 CF
Fragment	13 Frag
Misplaced Modifier	16 Mis
Overburdened Sentence	20 OBS
Parallel Structure	23 Paral
Periodic Structure	24 Per
Run-on (Fused) Sentence	30 RO
Sentence Beginning	31 SB
Subordination	36 Sub
Sentence Variety	37 SV